Biostatistics Casebook

BIOSTATISTICS CASEBOOK

Edited by

RUPERT G. MILLER, JR.

BRADLEY EFRON

BYRON WM. BROWN, JR.

LINCOLN E. MOSES

Stanford University

JOHN WILEY & SONS
New York • Chichester • Brisbane • Toronto • Singapore

Library of Congress Cataloging in Publication Data

Main entry under title:

Biostatistics casebook.

(Wiley series in probability and mathematical
statistics)
"A Wiley-Interscience publication."
1. Biometry. 2. Medical statistics. I. Miller,
Rupert G.
QH323.5.B563   574'.028   79-25405
ISBN 0-471-06258-8

Printed in the United States of America

10 9 8 7 6 5 4 3 2

CONTRIBUTORS

Byron Wm. Brown, Jr.
Division of Biostatistics
Stanford University Medical
  Center
Stanford, California 94305

Bradley Efron
Department of Statistics
Stanford University
Stanford, California 94305

Marie S. J. Hu
Division of Biostatistics
Stanford University Medical
  Center
Stanford, California 94305

John Hyde
Mathematical and Applied
  Statistics Branch
National Heart, Lung, and Blood
  Institute
Bethesda, Maryland 20205

Sue Leurgans
Department of Statistics
University of Wisconsin
Madison, Wisconsin 53706

Rupert G. Miller, Jr.
Department of Statistics
Stanford University
Stanford, California 94305

Lincoln E. Moses
Department of Statistics
Stanford University
Stanford, California 94305

Robert A. Wolfe
Department of Biostatistics
University of Michigan
Ann Arbor, Michigan 48109

Suzanna Wong
PPD Biostatistics
Abbott Laboratories
North Chicago, Illinois 60064

PREFACE

In 1975 we decided it would be worthwhile to record actual
case histories of biostatistical applications. The purpose was
to share some of our more interesting design and analysis problems with fellow statisticians and students. The cases were chosen to illustrate situations beyond standard textbook
methodology, and also to bring attention to some particularly
intriguing data sets. Most but not all of the cases came from
our regular consulting in the Division of Biostatistics at the
Stanford Medical School.

Volumes I (1976) and II (1978) of Biostatistics Casebook were
prepared as technical reports under our NIH research grant.
These proved to be very popular, and several people suggested
that it would be worth publishing them as a book for wider
circulation. We contacted Beatrice Shube, Editor at Wiley-
Interscience, and she was immediately interested in producing
them as a book. Volumes I and II appear herein as Parts I and
II.

Although it might have been possible to do so, no attempt was
made to refine the statistical analyses or substantially change
the presentations from their original forms. The thought was to
preserve the analyses as they were presented to the investigators
under the actual time constraints and not to delve into post hoc
academic exercises. Alternative analyses are possible in many

instances, and wherever feasible all the available data are presented so that the interested reader can try his/her more favor-
ite approach to the data for comparison purposes.

Rupert G. Miller, Jr.
Bradley Efron
Byron Wm. Brown, Jr.
Lincoln E. Moses

Stanford, California
April, 1980

ACKNOWLEDGMENTS

We would like to acknowledge the financial support of the National Institute of General Medical Sciences through Grant Number GM21215 in the preparation of these casebooks as technical reports.

We are very appreciative of the excellent work of Judi Davis, Karola Lof, Nancy Steege, and Beverly Weintraub in converting our questionable handwriting into typed technical reports, of the lovely figures drawn by Maria Jedd, of Jerry Halpern's assistance with computer printouts, and of Wanda Edminster's expert typing of the final manuscript for book production.

The Authors

CONTENTS

Biostatistics Casebook

PART ONE

PREDICTION ANALYSES FOR BINARY DATA

BYRON WM. BROWN, JR.

Medical Problem. Determining which of five preoperative variables are predictive of nodal involvement in cancer of the prostate, and how accurately prediction can be made.

Medical Investigator. Gordon Ray, Stanford University.

Statistical Procedures. Logistic regression analysis; contingency table analysis; Mantel-Haenszel test; discriminant
analysis.

MEDICAL BACKGROUND

When a patient is diagnosed as having cancer of the prostate,
an important question in deciding on treatment strategy for the
patient is whether or not the cancer has spread to the neighbor-
ing lymph nodes. The question is so critical in prognosis and
treatment that it is customary to operate on the patient (i.e.,
perform a laparotomy) for the sole purpose of examining the nodes
and removing tissue samples to examine under the microscope for
evidence of cancer. However, certain variables that can be measured without surgery are predictive of the nodal involvement; and
the purpose of the study presented here was to examine the data
for 53 prostate cancer patients receiving surgery, to determine
which of five preoperative variables are predictive of nodal
involvement, and how accurately the prediction can be made.
In particular, the medical investigator, Gordon Ray, was
interested in whether or not an elevated level of acid phosphatase in the blood serum would be of added value in the prediction

3

of whether or not the lymph nodes were affected, given the other four more generally used variables.

Table 1 presents the data. For each of the 53 cases, there are two quantitative variables, age at diagnosis and level of serum acid phosphatase (× 100), and three binary prediction variables, X-ray reading, pathology reading (grade) of a biopsy of the tumor obtained by needle before surgery, and a rough measure of the size and location of the tumor (stage) obtained by palpation with the fingers via the rectum. For the binary variables a value of one signifies a positive or more serious state and a zero (a blank in Table 1) denotes a negative or less serious finding. In addition, Table 1 presents the finding at surgery, a value of one denoting nodal involvement, and a value of zero (blank in Table 1) denoting no nodal involvement found at surgery. There were no missing values in this study.

CONTINGENCY TABLE ANALYSES

In the following preliminary analyses the quantitative variables were dichotomized, the division point for both age and acid phosphatase taken as 60 (i.e., < 60 vs. $\geq$ 60).

A small amount of clerical work revealed that a positive X ray is the best indicator for nodal involvement (see Table 2). Elevated acid phosphatase, using a cutoff value of 60, is not quite as effective. Note that the probabilities of false negatives and false positives for the two variables are estimated as follows:

$$P(-X \text{ ray})|+\text{nodes}) = 9/20, \qquad P(-\text{acid}|+\text{nodes}) = 4/20,$$

$$P(+X \text{ ray})|-\text{nodes}) = 4/33, \qquad P(+\text{acid}|-\text{nodes}) = 14/33 .$$

The probabilities of one error or the other, given the relative frequency of positive nodes seen in the sample of 53, are 13/53 and 18/53 for X-ray and acid phosphatase predictors, respectively.

Table 3 shows the result obtained if acid phosphatase and X ray are used jointly in predicting nodal involvement. Although the cells representing cases with both findings negative or both positive suggest fairly clear-cut prognoses, the intermediate cells present dilemmas, and the best prognostic rules (insisting on both findings positive versus one or both negative) has an estimated error rate of 13/55, no improvement over the use of X ray alone.

Now suppose that all five variables are used for prognosis. Table 4 presents the tabulation of patients with and without

TABLE 1. RAW DATA (SEE TEXT FOR DEFINITIONS OF THE VARIABLES)

Case Number	X Ray	Stage	Grade	Age	Acid	Nodes
1				66	48	
2				68	56	
3				66	50	
4				56	52	
5				58	50	
6				60	49	
7	1			65	46	
8	1			60	62	
9			1	50	56	1
10	1			49	55	
11				61	62	
12				58	71	
13				51	65	
14	1		1	67	67	1
15			1	67	47	
16				51	49	
17			1	56	50	
18				60	78	
19				52	83	
20				56	98	
21				67	52	
22				63	75	
23			1	59	99	1
24				64	187	
25	1			61	136	1
26				56	82	1
27		1	1	64	40	

5

TABLE 1 (Continued)

Case Number	X Ray	Stage	Grade	Age	Acid	Nodes
28		1		61	50	
29		1	1	64	50	
30		1		63	40	
31		1	1	52	55	
32		1	1	66	59	
33	1	1		58	48	1
34	1	1	1	57	51	1
35		1		65	49	1
36		1	1	65	48	
37	1	1	1	59	63	
38		1		61	102	
39		1		53	76	
40		1		67	95	
41		1	1	53	66	
42	1	1	1	65	84	1
43	1	1	1	50	81	1
44	1	1	1	60	76	1
45		1	1	45	70	1
46	1	1	1	56	78	1
47		1		46	70	1
48		1		67	67	1
49		1		63	82	1
50		1	1	57	67	1
51	1	1		51	72	1
52	1	1		64	89	1
53	1	1	1	68	126	1

TABLE 2. PREDICTION OF NODAL INVOLVEMENT BASED ON X RAY OR ACID PHOSPHATASE ALONE

Indicator		Nodal Involvement		
		0	1	
X ray	0	29	9	38
	1	4	11	15
		33	20	53
		0	1	
Acid phosphatase	< 60	19	4	23
	$\geq$ 60	14	16	30
		33	20	53

TABLE 3. PREDICTION OF NODAL INVOLVEMENT BASED ON BOTH X RAY AND SERUM ACID PHOSPHATASE LEVEL

Acid Phosphatase	X Ray	Nodal Involvement		
		0	1	
< 60	0	17	2	19
$\geq$ 60	0	12	7	19
< 60	1	2	2	4
$\geq$ 60	1	2	9	11
		33	20	53

TABLE 4. PREDICTION OF NODAL INVOLVEMENT BASED ON FIVE
PROGNOSTIC VARIABLES

| Prognostic Variable | | | | | Nodal Involvement | |
Age	Stage	Grade	X Ray	Acid	0	1
0	0	0	0	0	4	
0	0	0	1	0	1	
0	0	1	0	0	1	1
0	0	1	1	0		
0	1	0	0	0		
0	1	0	1	0		1
0	1	1	0	0	1	
0	1	1	1	0		1
1	0	0	0	0	3	
1	0	0	1	0	2	
1	0	1	0	0	1	
1	0	1	1	0		
1	1	0	0	0	2	1
1	1	0	1	0		
1	1	1	0	0	4	
1	1	1	1	0		
0	0	0	0	1	5	1
0	0	0	1	1	1	
0	0	1	0	1		1
0	0	1	1	1		
0	1	0	0	1	1	1
0	1	0	1	1		
0	1	1	0	1	1	2
0	1	1	1	1	1	5
1	0	0	0	1	3	

TABLE 4 (Continued)

	Prognastic Variable				Nodal Involvement	
Age	Stage	Grade	X Ray	Acid	0	1
1	0	0	1	1		1
1	0	1	0	1		
1	0	1	1	1		1
1	1	0	0	1	2	2
1	1	0	1	1		1
1	1	1	0	1		
1	1	1	1	1		1

nodal involvement for each of the $2^5 = 32$ cells formed by the five binary prognostic indicators. To answer the question concerning the usefulness of the acid phosphatase variable as an additional prognostic variable, the lines of Table 4 can be combined pairwise to form 16 fourfold tables, relating acid phosphatase to nodal involvement conditional on the other four variables. These tables are displayed in Table 5.

Of the 16 fourfold tables in Table 5, 10 have one or more zeros as marginal totals and hence do not offer any information on the conditional or partial correlation of acid level with nodal involvement. Of the remaining six, all but one, namely the eighth table, show a positive relationship. The sparseness of data suggests that this is slender evidence at best, however. The Mantel-Haenszel technique can be used to test the null hypothesis of no correlation, based on all the fourfold tables, by computing the expected values for the upper-left-hand cell frequency in each table, along with its hypergeometric variance, then summing and comparing with the observed sum. The tables with one or more zeros for marginal totals make no contribution and we have the following results:

TABLE 5.   FOURFOLD TABLES OBTAINED BY PAIRING LINES IN TABLE 4 TO
SHOW ACID PHOSPHATASE VS. NODAL INVOLVEMENT, HOLDING THE OTHER
FOUR PROGNOSTIC VARIABLES FIXED

Nodal Involvement

1)

	0	1	
< 60	4	0	4
≥ 60	5	1	6
	9	1	10

5)

0	1	
0	0	0
1	1	2
1	1	2

2)

1	0	1
1	0	1
2	0	2

6)

0	1	1
0	0	0
0	1	1

3)

1	1	2
0	1	1
1	2	3

7)

1	0	1
1	2	3
2	2	4

4)

0	0	0
0	0	0
0	0	0

8)

0	1	1
1	5	6
1	6	7

10

TABLE 5 (Continued)

Nodal Involvement

9)

3	0	3
3	0	3
6	0	6

13)

2	1	3
2	2	4
4	3	7

10)

2	0	2
0	1	1
2	1	3

14)

0	0	0
0	1	1
0	1	1

11)

1	0	1
0	0	0
1	0	1

15)

4	0	4
0	0	0
4	0	4

12)

0	0	0
0	1	1
0	1	1

16)

0	0	0
0	1	1
0	1	1

Fourfold Table Number	Upper-Left-Hand Cell Frequency		
	Observed	Expected[a]	Variance[a]
1	4	3.60	0.2400
3	1	0.67	0.2222
7	1	0.50	0.2500
8	0	0.14	0.1224
10	2	1.33	0.2222
13	2	1.71	0.4898
Sum	10	7.95	1.5466

[a]If the entries in the fourfold table are denoted by

$$\begin{array}{cc|c} a & b & r \\ c & d & s \\ \hline n & m & N \end{array}$$

the expected value and variance under hypergeometric assumptions are:

$$\text{Expected value} = n\left(\frac{r}{N}\right), \quad \text{Variance} = \frac{N-n}{N-1} \cdot n \cdot \frac{r}{N} \cdot \frac{s}{N} = \frac{m\,n\,r\,s}{(N-1)N^2} \ .$$

Thus the surplus of cases in the diagonal cell is 2.05. The test statistic, correcting for continuity, is

$$T = \frac{|10 - 7.95| - 1/2}{\sqrt{1.5466}} = 1.25 \ .$$

The normal tail area is P = 0.11, so the evidence that acid level has value as a prognostic indicator when the other four variables are known seems somewhat tenuous, at least by this analysis.

Another question that should be studied is the value of the five variables jointly in predicting nodal involvement. Examining the lines in Table 4, line by line, and classifying all cases on a line according to majority vote (predicting no nodal involvement in case of ties), we find that only 8 errors are made in classifying the 53 cases, a substantial reduction from the 13

errors using X ray alone. However, this error rate is estimated
by using the 53 cases as the basis for their own classification.
If each case is sequentially deleted from the 53 and classified
on the basis of the remaining 52 cases, an unbiased estimate of
the error rate achievable for a sample of 52 cases is obtained.
The result is 21 errors in 53 predictions, suggesting that the
use of the 5 variables in this way is very sensitive to the
occurrence of empty cells and cells with small frequency when the
total sample size is only 53 and that the estimate of 8/53 is
seriously biased in the optimistic direction.

MULTIPLE LOGISTIC REGRESSION ANALYSIS

The joint use of the five variables clearly requires some
smoothing in order to reduce the effect of sparse frequencies in
individual cells. One approach is through use of a model with a
modest number of parameters. If the multiple logistic model is
used, with no interactions, we have

$$\text{Prob } (+ \text{ nodes} | X_1 \cdots X_5) = [1 + \exp(-B_0 - \Sigma_i \, B_i X_i)]^{-1} .$$

This model can be fit by the maximum likelihood criterion, using
the variables singly or jointly in any combination. Further, the
quantitative variables, age and acid phosphatase level, need not
be dichotomized.

Table 6 presents the results of 10 logistic regression
analyses, employing single predictor variables, each of the first
four jointly with acid phosphatase level, and all five variables
jointly. The results in the table were all based on the dichotomized versions of the quantitative variables. Similar runs were
made using the quantitative values, with essentially the same
results, but no added statistical significance was achieved in
the latter runs. The results for the dichotomized runs are more
interesting in that the regression coefficients for the five variables are all in the same units (i.e., increment in the logit
associated with presence versus absence of the predictor
variable), and the results can be compared with the contingency
table results already obtained.

From Table 6 it can be seen that all variables except age are
statistically significant, including acid phosphatase, when used
singly, and that acid phosphatase retains its statistical significance when used jointly with any one or with all of the other
four variables. From the magnitude of the regression coefficients it can be seen that a positive X ray finding is associated

TABLE 6.   REGRESSION COEFFICIENTS AND STATISTICAL SIGNIFICANCE
FOR THE LOGISTIC REGRESSION ANALYSIS

| Computer Run | Regression Coefficients ($\hat{B}_i$) | | | | |
	X Ray	Stage	Grade	Age	Acid
1				-0.63	
2		1.66[a]			
3			1.39[b]		
4	2.18[a]				
5					1.69[a]
6				-0.54	1.66[b]
7		1.67[b]			1.71[b]
8			1.65[b]		1.93[a]
9	2.07[a]				1.57[b]
10	1.97[b]	1.35	0.76	-0.62	1.63[b]

[a]Significant at 1% level.

[b]Significant at 5% level, using asymptotic normal test.

with the largest increment in the logit or log odds ratio
$\ln[P/(1-P)]$, that acid phosphatase is as important as any of the
others, and that age is negatively correlated with nodal involvement, though the regression coefficient for age is not statistically significant and the size is small.

The constant term for computer run 10 was estimated as -2.82,
and Table 7 presents the logistic probability estimates for various combinations of the predictor variables, using this constant
and the regression coefficients given for run 10 in Table 6.
Table 7 shows the minimum probability to be 3%, rising to 14% if
acid phosphatase is elevated and all other predictor variables
remain negative. If all findings are positive, the probability
of nodal involvement is estimated at 95%, decreasing to 78% if
all are positive except acid phosphatase level. It should be

TABLE 7. PROBABILITIES OF NODAL INVOLVEMENT FOR VARIOUS COMBINATIONS OF PREDICTOR VARIABLES, BASED ON A LOGISTIC REGRESSION ANALYSIS OF ALL DATA

Variable	Coefficient	Values of Prediction Variables			
Constant	-2.82	1	1	1	1
X ray	1.97	0	0	1	1
Stage	1.35	0	0	1	1
Grade	0.76	0	0	1	1
Age	-0.62	1	1	0	0
Acid	1.63	0	1	0	1
Pr(Nodal involvement)		0.03	0.14	0.78	0.95

remarked again here that age under 60 is regarded as a positive finding.

With regard to the errors of prediction, when nodal involvement is predicted for the 53 cases, based on the fitted model, the optimal cutoff for the sample of 53 cases was found by computing the estimated logistic probability for each case, ordering and listing them, and searching by trial and error. The optimal point was .35, and the proportions of false positives and negatives were 3/20 and 6/33, for a total error rate of 9/53. The logistic analysis was not repeated 53 times, each time with one case deleted, to obtain an unbiased estimate of the error, but it is not likely that the estimate would be changed much, in light of the small number of parameters estimated.

Thus the logistic analysis seems much more persuasive in indicating the value of acid phosphatase as a predictor variable, even conditional on the other variables, and the prognostic rules based on the logistic function seem reliable and useful.

The model used did not allow for interactions among the prognostic variables and nodal involvement. To obtain some assurance that a useful interaction had not been missed, a stepwise logistic regression analysis was done, restricting the search to the first-order interactions only. Thus the variables allowed were the X_i themselves and the 10 cross products $X_i X_j$. Table 8 presents some results of the stepwise analysis. The variables were added according to the criterion of maximum partial correlation between the remaining variables and the residuals, based on the

TABLE 8.   REGRESSION COEFFICIENTS AND STATISTICAL SIGNIFICANCE
FOR STEPWISE LOGISTIC REGRESSION, USING FIVE BINARY PROGNOSTIC
VARIABLES AND THEIR FIRST-ORDER INTERACTIONS

| Step Number | Variables Chosen (from 5 + 10 Interactions) | | | | | χ^2 for New Variable |
	X Ray	Stage	Stage × Grade	Acid	Grade	
1	2.18[a]					11.25
2	2.12[a]	1.59[b]				5.65
3	2.66[a]	1.65[b]	−3.40[b]			4.74
4	2.57[a]	1.48[b]	−3.59[b]	1.63[b]		4.47
5	2.57[a]	1.77	−4.73[b]	1.97[b]	1.69	5.91

[a]Statistically significant at the 1% level.

[b]Statistically significant at the 5% level, based on the usual asymptotically normal test statistic.

model fitted at the previous step. As before, X ray was a strong
predictor throughout all steps; stage was added next in the stepwise procedure. The interaction of stage with grade was next and
the coefficient had a negative sign, indicating that a positive
finding for stage (or grade) is important if the finding for the
other is not positive, but if both stage and grade are positive,
then their joint contribution to the probability of nodal
involvement is overstated by the sum of their separate contributions. This analysis cannot be taken too seriously, however,
since the increase in likelihood achieved by adding this new variable is only of borderline statistical significance, as measured
by the asymptotic chi-squared test, shown in the rightmost column
of Table 8. Acid phosphatase and grade were the next variables
chosen, their added contributions to the likelihood being unim-
pressive though their regression coefficients are large enough to
be of importance.

The more complex model, employing an interaction and achieved
by stepwise analysis, makes eight errors at the optimal cutoff
point of .44, which is no important improvement over the nine
errors made with the simple model, without interactions and estimated without a stepwise choice of variables. Since the gain is
not great and, in fact, the logistic regression estimated by the

stepwise procedure might be expected to be rather unstable with optimistic bias in the error rate due to the large number of predictor variables used (15), it would seem prudent to use the simpler model and keep the interaction model in mind for further checking on additional patients.

The instability of the variables and estimates chosen by the stepwise analysis with interactions could have been checked by running the program 53 times, with one patient omitted at each run and classified on the basis of the remaining 52. The cost of this was too high. Instead the 32 cells shown in Table 4 were grouped into the eight cells suggested by run 4 shown in Table 8 (i.e., acid positive, negative; X ray positive, negative; stage and grade both positive vs. one or both negative) and the error rates were tabulated by classifying within these cell groups. The error rate estimated was 10/53 (compared with 8/53 based on the logistic run itself), but when individual patients were deleted sequentially, and each classified only on the basis of his group neighbors, the unbiased error rate estimate doubled (21/53), clearly demonstrating the instability of this grouping of cells for prognosis.

CONCLUSIONS

This study will continue with the study of more prostate cancer patients as they are diagnosed at the Stanford University Medical Center. However, the tentative conclusion based on the data thus far is that acid phosphatase is a useful prognostic indicator, even after the other variables (X ray, stage, age, grade) are known; that it can substantially alter the estimated probability of nodal involvement after one knows the other indicators; and that the logistic model with no interactions seems the best prognostic approach for the moment. In certain cases the logistic model may provide a probability estimate so close to zero or 1 as to obviate the need for surgery, especially in the patient considered a poor risk with regard to the surgical procedure.

ACKNOWLEDGMENTS

We are grateful to Gordon Ray for his permission to use his data in this consulting note. Jerry Halpern of the Division of Biostatistics did the computer programming and analysis.

REFERENCES

For the Mantel-Haenszel method of pooling results from 2×2 tables, see their paper titled "Statistical aspects of the analysis of data from retrospective studies of disease," J. Nat. Cancer Inst., 22, 719-748. For a discussion of the use of the logistic model in the analysis of binary data, see D. R. Cox (1970). The Analysis of Binary Data. London: Methuen.

RANDOMIZING AND BALANCING A COMPLICATED SEQUENTIAL EXPERIMENT

BRADLEY EFRON

<u>Medical Problem</u>. Investigating the effectiveness of various treatments for ovarian carcinoma.

<u>Medical Investigators</u>. Zvi Fuks and Myron Turbow, Stanford University.

<u>Statistical Procedures</u>. Randomization with forced balancing.

MEDICAL BACKGROUND

The investigators wish to explore the effectiveness of various radiation and chemical treatments for ovarian carcinoma. The subjects all will have first received the standard surgical treatment appropriate for their state of the disease. The experiment will actually consist of seven subexperiments carried out simultaneously, each of which will compare two or three of the possible treatments. This is necessary because some of the treatments, such as "no further therapy," are appropriate only for patients who have been diagnosed to have a low (less advanced) stage of the disease.

Figure 1 shows the experimental design. The seven protocols, that is, subexperiments, are described in terms of the stage of the disease, the degree of cell differentiation observed, and the results of a postoperative lymphangiograph (LAG). Each protocol has a less favorable prognosis than its predecessor. The figure shows that eight treatments will be investigated, those being various combinations of no treatment, pelvic radiation, whole abdominal radiation, postaortic radiation, and two combination

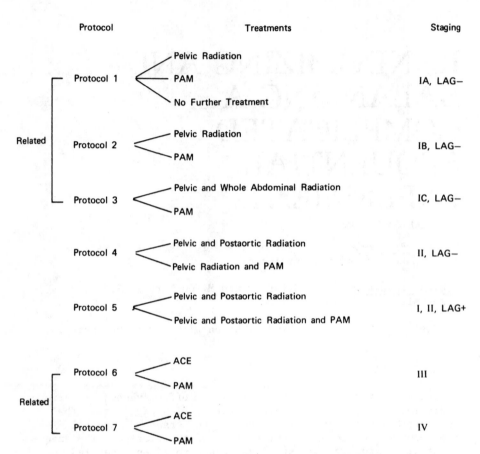

Figure 1. The seven protocols and eight possible treatments. I, II, III, and IV indicate standard FIGO staging; A, B, and C indicate well, moderately, and poorly differentiated cells. LAG- and LAG+ indicate negative or positive lymphangiograph. Protocols 1, 2, and 3 are associated with each other in the sense that we can reasonably expect to pool at least some of their data in the eventual analysis; likewise protocols 6 and 7.

chemotherapies, PAM and ACE. See Bagley, et al. (1972), Hanks and Bagshaw (1969), Kraut et al. (1972), Parker et al. (1974), and Smith and Rutledge (1970).

Several observed covariates affect the prognosis of the disease. The most importnat of those is the tumor histology, which for our purposes takes on one of six values: (1) serous, (2) mucinous, (3) endometroid, (4) clear cell, (5) undifferentiated, and (6) other. Table 1 shows the frequencies of these six types by stage for 151 cases previously treated at Stanford.

THE DESIGN PROBLEM

Subjects enter the study sequentially at a rate of about one per week. They must be assigned to a treatment immediately. Complete randomization would make the assignments by means of a random device operating independently in each case. For example, a new patient in protocol 2 would be assigned either to pelvic radiation or PAM by, essentially, the flip of a fair coin, while in protocol 1 one of three possible treatments would be selected with equal probability. Complete randomization protects the experiment against latent variable bias and selection bias, and can also be used as a basis for statistical inference (Efron, 1971).

Unfortunately, in an experiment with seven protocols and six histologies, complete randomization almost guarantees that there will be some unpleasant imbalances in the allocation of experimental units. After 50 assignments, about one year of the experiment, it might turn out that all of the protocol 1 PAM recipients were in the first three histologies, while all the protocol 1 pelvic radiation recipients were in the last three histologies. Simple probability calculations show that at least one such disaster is expected to happen in the situation described.

We could minimize such imbalances simply by assigning every other patient in each protocol-histology combination to alternate treatments (with an obvious modification for protocol 1). Such designs have a well-deserved bad reputation for vulnerability to latent variables and selection bias.

This problem is discussed in Efron (1971) for the simpler situation where there is only one protocol under consideration, and no important covariates. A compromise between complete randomization and the alternation method is proposed. Suppose that just before the n-th patient is to be assigned, n_1 patients have previously been assigned to treatment 1 and n_2 patients to treatment 2. The, if $n_1 > n_2$, the next assignment is made to treatment 2 with probability p and to treatment 1 with probability q, where p is some number greater than 1/2, say 2/3 or 3/4. If $n_2 > n_1$,

TABLE 1. THE FREQUENCY BY STAGE AND HISTOLOGICAL TYPE OF 151 PREVIOUSLY OBSERVED OVARIAN CARCINOMA PATIENTS

	Histology						
Stage	1. Serous	2. Mucinous	3. Endometroid	4. Clear Cell	5. Undifferentiated	6. Other	Total
I	8	0	8	7	0	7	30(20%)
II	20	3	9	1	10	19	62(41%)
III	28	0	3	0	11	13	55(36%)
IV	1	0	0	0	2	1	4(3%)
Total	57(38%)	3(2%)	20(13%)	8(5%)	23(15%)	40(27%)	151(100%)

22

then the probabilities are reversed.   (If $n_1 = n_2$ the treatments are assigned with equal probabilities.)   In other words, the assignments are made by independent flips of a biased coin, where the bias is always selected to force the experiment in the direction of balance.   It is shown in Efron (1971) that such biased coin designs tend to produce well-balanced experiments, while retaining most of the virtues of complete randomization.

A BIASED COIN DESIGN FOR THIS SITUATION

   In the simple situation considered in Efron (1971) it is always clear which treatment should be assigned next in order to improve the balance of the experiment.   Such is not the case for the problem at hand.   Consider the hypothetical situation described in Table 2.   A new patient has arrived in protocol 6, histology 2.   The previous assignments are as shown.   For example, no previous assignments have been made to ACE in protocol 6, histology 2, while one previous assignment has been made to PAM.   By itself this would suggest biasing the current assignment to ACE.   However, protocol 6 as a whole has two previous assignments to ACE and only one to PAM, so the reverse bias might be called for.   Likewise, the total for histology 2 over the associated protocols 6 and 7 shows an imbalance in favor of ACE.
   Of course, it would be ideal to balance assignments perfectly within each protocol-histology category, but with 42 combinations this would be an unrealistic goal during the first 100 or so assignments.   It is more reasonable to try to balance in the protocol and histology marginals (remembering that the latter is taken only over the associated protocols since it is felt that only the data from these can be combined statistically).   The following assignment scheme aims to do just that, while still trying for balance within individual protocol-histology categories whenever that is possible.
   Let $n_{ij}(k)$ be the number of patients previously assigned to treatment k in protocol i, histology j.   Here i = 1,2,...,7 and j = 1,2,...,6.   For protocol 1, k = 1,2,3, while for the other protocols k = 1,2.   In the hypothetical example of Table 2, $n_{6,2}(1) = 0$, $n_{6,2}(2) = 1$, and so on.   Also let $n_{i+}(k) \equiv \Sigma_j n_{ij}(k)$, the number of previous assignments to treatment k in protocol i, and $n_{+j}(k) = \Sigma_i n_{ij}(k)$, the sum (over associated protocols) for previous assignments of treatment k to histology j.   In Table 2 $n_{6+}(1) = 2$, $n_{+2}(1) = 2$.

TABLE 2.   A HYPOTHETICAL SITUATION IN WHICH THE ASSIGNMENTS
PREVIOUSLY MADE TO ACE AND PAM, RESPECTIVELY, ARE AS SHOWN.
A NEW PATIENT ARRIVES IN PROTOCOL 6, HISTOLOGY 2.   WHICH
ASSIGNMENT WOULD WE PREFER HER TO HAVE?

		Histology						
		1	2	3	4	5	6	Total
Protocol	6	0-0	0-1	0-0	1-0	1-0	0-0	2-1
	7	0-1	2-0	0-0	0-1	1-0	0-1	3-3
Total		0-1	2-1	0-0	1-1	2-0	0-1	5-4

The idea of the assignment algorithm is this: given a new
patient in protocol i, histology j, a single composite score is
calculated for each treatment k from $n_{ij}(k)$, $n_{i+}(k)$, and $n_{+j}(k)$.
The treatment with the smallest score is favored in making the
assignments. The scoring function is

$$S(k) = t[n_{ij}(k)]t^{1.5}[n_{i+}(k)]t^{0.5}[n_{+j}(k)] ,$$

where $t[n]$ is the peculiar function described in Table 3.
In protocols 2-7 the treatment with the smaller value of $S(k)$
was assigned next with probability 3/4, while that with the
larger value was assigned with probability 1/4. In case of a tie
they were assigned with equal probabilities. A more complicated
rule, which won't be described here, was used for the three-way

TABLE 3.   THE FUNCTION $t[n]$ USED IN THE SEQUENTIAL ASSIGNMENT
PROCESS

n	0	1	2	3	4	5	6	7	8	9
t[n]	1	4	10	15	20	24	27	30	33	36

n	10	11	12	13	14	15	16	17	...
t[n]	38	40	42	44	46	47	48	49	...

choice in protocol 1. In the example of Table 2,
$$S(1) = t[0]t^{1.5}[2]t^{0.5}[2] = 100, \quad S(2) = t[1]t^{1.5}[1]t^{0.5}[1] = 64,$$
so the next assignment is to treatment 2 with probability 3/4.

Figure 2 shows the ratio $t[n(1)]/t[n(2)]$ for various combinations $(n(1)),n(2))$. Notice for example that the ratio for $(1,0)$ is greater than that for $(3,1)$ but less than that for $(4,1)$. This means that the scoring function $S(k)$ tends to treat an imbalance $(1,0)$ less seriously than an imbalance $(4,1)$, but more seriously than an imbalance $(3,1)$. The function t was chosen to given approximately the ordering of pairs shown in Figure 2. This reflected the author's own ordering of how urgently an $(n(1),n(2))$ combination requires balancing. To put it another way, the author would feel worse about going from $(1,0)$ to $(2,0)$ [rather than to $(1,1)$] than about going from $(3,1)$ to $(4,1)$. Once the ordering was made there was surprisingly little leeway in choosing the function t.

The powers 1.5 and 0.5 that appear in the formula for $S(k)$ were selected to give priority to balancing over protocol marginals rather than histology marginals, since it was not clear how useful the "associated protocols" were actually going to be in the eventual analysis. The choice was also based on some Monte Carlo studies, which showed good balancing performance both for small and large numbers of assignments. These are described next.

MONTE CARLO STUDY

In the Monte Carlo study it was supposed that patients enter the study randomly with the probability of being in protocol i, histology j being $P_{prot}(i) \cdot P_{hist}(j)$:

i	1	2	3	4	5	6	7
$P_{prot}(i)$	0.05	0.05	0.05	0.20	0.25	0.37	0.03

j	1	2	3	4	5	6
$P_{hist}(j)$	0.38	0.02	0.13	0.05	0.15	0.27

These two probability distributions are based on the marginal probabilities in Table 1. The independence assumption is obviously unrealistic, but was deemed good enough for this exercise.

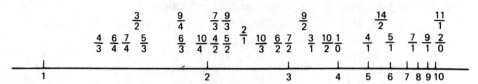

Figure 2.   The ratio $t[n_1]/t[n_2]$ is plotted for various choices of n_1 and n_2.   For example, $t[1]/t[0] = 4$.

The sequential assignment scheme described above was carried out with the results shown in Table 4.   The results are quite good, especially early in the experiment when complete randomization would often produce unpleasant imbalances.   Obviously the method described here has many ad hoc features, but in a complicated sequential design problem like the one at hand it does not seem to have any serious competitors.   Pocock and Simon (1975) have proposed another variant of the scheme in Efron (1971), actually quite similar to the one considered here, which they show to have good performance properties.

TABLE 4. MONTE CARLO STUDY OF THE SEQUENTIAL ASSIGNMENT SCHEME AFTER VARIOUS NUMBERS n OF PATIENTS HAVE BEEN ASSIGNED. EMPTY CELLS INDICATE NO ASSIGNMENTS MADE. FOR EXAMPLE, AFTER 80 TRIALS, IN THE FOURTH PROTOCOL AND SIXTH HISTOLOGY, 2 PATIENTS HAD BEEN ASSIGNED TO THE FIRST TREATMENT (ACE) AND 3 TO THE SECOND (PAM)

n = 20

	1	2	3	4	5	6	Total
1							0,0,0
2	1,0				0,1		1,1
3							0,0
4	3,1				1,0		5,2
5	2,1						2,1
6	3,2						4,3
7					1,0		0,1

n = 40

	1	2	3	4	5	6	Total
1							0,0,0
2	1,0				0,1		1,1
3			1,1	1,0			2,1
4	3,2		1,0		1,0		6,4
5	2,4		1,1		1,0		3,6
6	5,5				0,1		7,8
7					1,0		1,0

TABLE 4. (Continued)

n = 80

	1	2	3	4	5	6	Total
1						1,0	1,0,0
2	1,1				0,1	0,1	1,3
3			1,1	1,0			2,1
4	3,4		2,1	1,0	2,0	2,3	10,8
5	5,7		2,1		2,3	1,1	10,12
6	10,8				1,3	4,5	15,16
7					1,0		1,0

n = 180

	1	2	3	4	5	6	Total
1	0,0,1		1,0,0		1,0,0	1,1,0	3,1,1
2	2,2		1,1		0,1	2,2	5,6
3	0,1		3,2	1,0	1,1	0,1	5,5
4	6,7		4,4	2,1	4,1	2,5	18,18
5	12,12		5,3	0,1	4,3	3,3	24,22
6	15,14	1,1	2,2	0,1	7,6	9,9	34,33
7			0,1		1,1	0,2	1,4

TABLE 4. (Continued)

n = 280

	1	2	3	4	5	6	Total
1	0,1,1		2,1,0		1,0,0	1,2,1	4,4,2
2	2,2		1,2	1,0	0,2	2,2	6,8
3	2,2		3,2	1,0	1,1	1,2	8,7
4	13,12		8,9	3,2	5,4	6,6	34,33
5	12,15	0,1	8,7	1,1	6,6	7,7	34,37
6	19,18	1,1	4,2	1,2	10,11	12,12	47,46
7	2,0		0,1		2,1	1,3	5,5

n = 1080

	1	2	3	4	5	6	Total
1	6,6,6		3,3,3	0,0,1	1,2,0	5,3,5	15,14,15
2	13,13	0,1	2,4	2,1	2,4	13,10	32,33
3	6,6		7,6	2,1	5,2	8,8	28,23
4	44,43	1,3	18,18	4,2	16,14	28,29	111,109
5	55,56	1,2	22,22	7,6	21,21	33,32	139,139
6	71,69	5,5	28,28	9,8	30,32	49,49	192,191
7	8,7	1,0	1,2	1,2	5,4	3,5	19,20

29

REFERENCES

Bagley, C. M., Jr., R. C. Young, G. P. Canellos, and B. T. De Vita, Jr. (1972). The treatment of ovarian carcinoma: Possibilities for progress. New Engl. J. Med., 287, 856-862.

Efron, B. (1971). Forcing a sequential experiment to be balanced. Biometrika, 58, 403-417.

Hanks, G. E. and M. A. Bagshaw (1969). Megavoltage radiation therapy and lymphangioraphyin ovarian cancer. Radiology, 93, 649-654.

Kraut, J. W., H. S. Kaplan, and M. A. Bagshaw (1972). Combined fractionated isotopic and external irradiation of the liver in Hodgkin's disease. Cancer, 30, 34-46.

Parker, B. R., R. A. Castellino, Z. Y. Fuks, and M. A. Bagshaw (1974). The role of lymphography in patients with ovarian cancer. Cancer, 34, 106-110.

Pocock, J. J. and R. Simon (1975). Sequential treatment assignment with balancing for prognostic factors in the controlled clinical trial. Biometrics, 31, 103-117.

Smith, J. P. and R. Rutledge (1970). Chemotherapy in the treatment of cancer of the ovary. Am. J. Obstet. Gynecol., 107, 691-703.

SURVIVAL ANALYSIS WITH INCOMPLETE OBSERVATIONS

JOHN HYDE

Medical Problem. Assess the impact of the Channing House medical program on survival.

Medical Investigator. Walter Bortz, Palo Alto Medical Clinic.

Statistical Procedures. Kaplan-Meier curves; one-sample hazard rate statistics.

MEDICAL BACKGROUND

Channing House is a retirement center located in Palo Alto, California. Its distinctive feature is that the residents are members of a total health insurance program. They are charged a fixed monthly fee which covers any medical care they require, with the single exception of physical therapy. Residents, therefore, should feel no financial inhibitions in seeking medical attention. A recent economic study by Anne Scitovsky and Nelda Snyder of the Palo Alto Medical Research Foundation shows that Channing House residents make greater use of medical facilities than senior citizens on conventional health insurance plans.

One might be led to ask the question: Does the Channing House program offer better health to the aged? This is a difficult question to answer because health is a difficult quantity to define. It could involve such things as life expectancy, extent of hospitalization, and amount of suffering. To be specific in a way that will lead to a tractable problem, let us consider whether Channing House helps to lengthen lifetimes.

Channing House is not run as a controlled experiment, so there is no readily available control population with which to make comparisons. It is unlikely that Channing House residents are typical of the general population since the expenses tend to exclude members of the lower socioeconomic brackets. Also, a fair portion of the residents are retired professors or their spouses. This is a group that has a history of greater life expectancy. Fortunately, some insurance companies have life tables reflecting the reduced mortality rate of the special groups they insure. In particular, Teachers Insurance Annuity Association (TIAA) insures a large proportion of university faculty. With the kind cooperation of Herbert Weiss at TIAA, we were able to obtain a copy of their annuity tables through the age of 110. We also obtained a copy of the 1970 Public Health Service (PHS) life tables through age 85. The latter should represent a baseline mortality rate of the general population.

The question of whether Channing House improves health has been refined to the question of whether the mortality rate at Channing House is less than that among the population used in constructing TIAA's annuity tables. No claim is made that the questions are equivalent or that the answers should be the same, but it is hoped that by attacking the specific problem we can learn something about the more general one.

THE DATA

Channing House began operation in January of 1964 and the closing date for this study was July 1, 1975. During most of that period there were about 270 people living in Channing House. There are records of 464 residents who were at Channing House sometime during the observation period. We obtained data on the sex, birthdate, entry date, and date of either death or withdrawal from Channing House. All dates were collected only by month and year since exact dates were scarce. For two cases the year of birth was unknown. These cases were excluded from the analysis. In the remaining 462 cases the years for each event were known, and June, being in the middle of the year, was used where the exact month was not known. From this information we computed the age in months upon entering Channing House and the age in months at the time of death or withdrawal. These data are shown in the appendixes and some summary statistics are given in Table 1.

TABLE 1.   SOME SUMMARY STATISTICS FROM THE CHANNING HOUSE DATA

	Men	Women	Total
Total	97	365	462
Died	46	130	176
Withdrew	5	22	27
Alive 7/1/75	46	213	259
Birth month missing	29	53	82
Entry month missing	7	18	25
Exit month missing	1	5	6

SOME ASSUMPTIONS

In the analysis that follows we will be assuming that if we
know the age of a person who entered Channing House, then knowing
the person's entry age will provide no additional information
about prospects for survival. Similarly, we assume that if we
know the age of someone who withdrew, then knowing the age at
withdrawal provides no more information about survival prospects
than had the person stayed. These assumptions would probably be
untenable if subjected to careful scrutiny. In fact, one
requirement for admission is that the applicant be capable of
self-care.

However, there is reason to believe the assumptions are
acceptable approximations. Many residents report that they
entered for reasons of financial security more than for reasons
of health. Those who withdraw do so often because they do not
feel "ready" for Channing House. An examination of survival
curves for selected ages of entry showed no unusual behavior in
the first years following entry, indicating that the entrance
requirement might not have much effect on the mortality rate.

STATISTICAL ANALYSIS

Using the technique of Kaplan and Meier, survival curves were
computed and compared with those from the TIAA and PHS data.
This was done separately for each sex. The curves are shown in
Figures 1 and 2. The Channing House curves were normalized to
show 100% survival at the youngest entry age for each sex. The
TIAA and PHS curves were made to match those of Channing House at
age 75 for men and age 70 for women. These ages were chosen to

SURVIVAL CURVES OF MEN

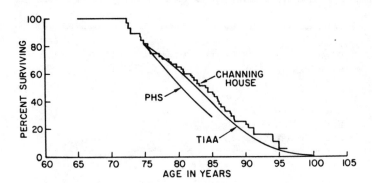

Figure 1.   Survival curves of men.

give large ranges for comparison while also avoiding unstable
portions of the curves.   The curves tend to indicate that members
of the Channing House group survive longer than the PHS data predict but that they do about as well as the TIAA data predict.

These observations can be tested by examining the excess of
observed deaths over the number we would expect if we were making
observations from the PHS or TIAA distributions.   The expectation

SURVIVAL CURVES OF WOMEN

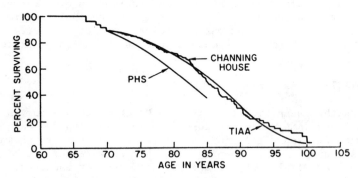

Figure 2.   Survival curves of women.

is found by summing the mortality rate over all of the observed lifetimes. Properly normalized, this excess has a limiting normal distribution.

To be precise, let p_i be the conditional probability of dying in the i-th month given survival until the beginning of that month. These values are determined from the life tables. Let ν_j be the age at which person j entered Channing House, let λ_j be the age upon leaving (due to death, withdrawal, or closing of the study), and let δ_j be 1 if the person died in Channing House, 0 otherwise. The total number of deaths observed in a sample of size n is $\Sigma_{j=1}^n \delta_j$ while the expected number is $\Sigma_{j=1}^n \Sigma_{i=\nu_j}^{\lambda_j} p_i$. An unbiased estimator of the variance of the difference between these quantities is $\Sigma_{j=1}^n \Sigma_{i=\nu_j}^{\lambda_j} p_i(1-p_i)$. It can be shown that if the p_i's come from the true distribution of the lifetimes, and if the assumptions of the previous section hold, then the normalized excess,

$$\frac{\sum_{j=1}^{n}\left(\delta_j - \sum_{i=\nu_j}^{\lambda_j} p_i\right)}{\left[\sum_{j=1}^{n} \sum_{i=\nu_j}^{\lambda_j} p_i(1-p_i)\right]^{\frac{1}{2}}}$$

has a standard normal limiting distribution.

The results of these tests are shown in Table 2. Since the PHS data stop at the age of 85 years, people who entered at 85 or later were not included in the PHS computations. This left 87 men and 348 women. Those who lived past 85 were treated as if they had withdrawn at 85.

CONCLUSIONS

The tests show that the Channing House mortality rate is substantially less than the PHS rate for both sexes. However, it is not significantly different from the TIAA rate for either sex.

TABLE 2. TEST RESULTS

	Men			
Group	Deaths Observed	Deaths Expected	Estimated Variance	Normalized Excess
PHS	28	44.3	43.4	−2.48
TIAA	46	50.3	47.8	−0.63
	Women			
PHS	83	120.0	119.2	−3.39
TIAA	130	125.5	119.9	+0.41

In fact, the rate for women at Channing House shows a tendency to be higher than the TIAA rate. In answer to the question posed at the end of the first section, we have to say no, the mortality rate of the Channing House group does not appear to be less than the rate for the population used in constructing TIAA's tables.

REFERENCES

Bortz, W.M., M. Kern, and J. Hyde (1977). A prepaid medical plan. J. Am. Med. Assoc., 238, 1269–1271.

Hyde, J. (1977). Testing survival under right censoring and left truncation. Biometrika, 64, 225–230.

Kaplan, E.L. and P. Meier (1958). Nonparametric estimation from incomplete observations. J. Am. Stat. Assoc., 53, 457–481.

Scitovsky, A.A. and N.M. Snyder (1975). Medical care use by a group of fully insured aged: A case study. Pub. No. 75-3129, Washington, D.C.: Health Research Administration, Department of Health, Education and Welfare, Public Health Service.

APPENDIX 1: CHANNING HOUSE MEN

LOSS (0) OR DEATH (1)	AGE IN MO. AT ENTRY	AGE IN MO. AT EXIT	MO. SPENT IN STUDY
1	782	909	127
1	1020	1128	108
1	856	969	113
1	915	957	42
1	863	983	120
1	906	1012	106
1	955	1055	100
1	943	1025	82
1	943	1043	100
1	837	945	108
1	966	1009	43
1	936	971	35
1	919	1033	114
1	852	869	17
1	1073	1139	66
1	925	1036	111
1	967	1085	118
0	806	943	137
0	969	1001	32
0	923	1060	137
0	865	1002	137
0	953	1031	78
0	871	945	74
0	982	1006	24
0	883	959	76
0	817	843	26
0	875	1012	137
0	821	956	135
0	936	1073	137
0	971	1107	136
0	830	940	110
0	885	911	26
0	894	1031	137
0	893	996	103
0	866	895	29
0	878	1015	137
0	820	957	137
0	1007	1043	36
0	879	1016	137
0	956	1093	137
1	854	989	135
0	890	1027	137
0	1041	1044	3
0	978	1005	27
0	836	973	137
0	938	1064	126
0	886	1023	137
0	876	1013	137
0	955	977	22
0	823	960	137

37

LOSS (0) OR DEATH (1)	AGE IN MO. AT ENTRY	AGE IN MO. AT EXIT	MO. SPENT IN STUDY
0	960	1047	87
0	843	943	100
0	856	951	95
0	847	984	137
0	1027	1058	31
0	988	1045	57
0	953	953	0
0	978	1018	40
0	981	1118	137
0	926	970	44
0	1036	1070	34
0	1016	1153	137
0	969	1106	137
0	900	936	36
0	898	906	8
0	846	866	20
1	964	1029	65
1	984	1053	69
1	1046	1080	34
1	871	872	1
1	847	893	46
0	962	966	4
1	853	894	41
1	967	985	18
1	1063	1094	31
1	856	927	71
1	865	948	83
1	1051	1059	8
1	1010	1012	2
1	878	911	33
1	1021	1094	73
0	959	972	13
1	921	993	72
1	836	876	40
1	919	993	74
1	751	777	26
1	906	966	60
1	835	907	72
1	946	1031	85
1	759	781	22
0	909	914	5
1	962	998	36
1	984	1022	38
1	891	932	41
1	835	898	63
1	1039	1060	21
1	1010	1044	34

```
             APPENDIX 2:  CHANNING HOUSE WOMEN
 LOSS (0) OR    AGE IN MO.      AGE IN MO.      MO. SPENT
 DEATH (1)      AT ENTRY        AT EXIT         IN STUDY
     1            1042            1172            130
     1             921            1040            119
     1             885            1003            118
     1             901            1018            117
     1             808             932            124
     1             915            1004             89
     1             901            1023            122
     1             852             908             56
     1             828             868             40
     1             968             990             22
     1             936            1033             97
     1             977            1056             79
     1             929             999             70
     1             936            1064            128
     1            1016            1122            106
     1             910            1020            110
     1            1140            1200             60
     1            1015            1056             41
     1             850             940             90
     1             895             996            101
     1             854             969            115
     1             957            1089            132
     1            1013            1115            102
     1            1073            1192            119
     1             976            1085            109
     1             872             976            104
     1            1027            1142            115
     1            1071            1200            129
     1             919            1017             98
     1             894            1006            112
     1             885            1012            127
     1             889            1000            111
     1             887            1012            125
     1             920            1040            120
     1            1015            1024              9
     1             942            1070            128
     1             924            1014             90
     1             883             996            113
     1             930             944             14
     1             956            1085            129
     1             886             994            108
     1             987            1097            110
     1             883             966             83
     1             837             948            111
     1             958            1029             71
     1             850             963            113
     1             890             905             15
     1             847             970            123
     1             919            1015             96
     1             748             804             56
```

39

LOSS (0) OR DEATH (1)	AGE IN MO. AT ENTRY	AGE IN MO. AT EXIT	MO. SPENT IN STUDY
1	934	1041	107
1	895	991	96
1	874	982	108
1	877	989	112
1	900	959	59
1	957	1084	127
1	1013	1131	118
1	967	1068	101
0	904	919	15
0	829	848	19
0	842	979	137
0	802	876	74
0	840	938	98
0	792	929	137
0	837	848	11
0	941	1006	65
0	746	804	58
0	834	932	98
0	865	932	67
0	828	965	137
0	894	1011	117
0	874	1011	137
0	917	1054	137
0	993	1028	35
0	918	1055	137
0	818	955	137
0	984	1019	35
0	1002	1010	8
0	857	994	137
0	827	836	9
0	883	1020	137
0	1008	1042	34
0	954	1091	137
0	905	1042	137
0	838	975	137
0	934	946	12
0	872	940	68
0	918	922	4
0	844	981	137
0	805	928	123
0	922	1059	137
0	821	958	137
0	838	975	137
0	934	961	27
0	886	1023	137
0	934	1023	89
0	878	1010	132
0	935	1072	137
0	799	825	26
0	849	952	103

APPENDIX 2: CHANNING HOUSE WOMEN (CONT.)

LOSS (0) OR DEATH (1)	AGE IN MO. AT ENTRY	AGE IN MO. AT EXIT	MO. SPENT IN STUDY
0	920	953	33
1	948	998	50
0	968	1105	137
0	908	996	88
0	828	965	137
0	897	1034	137
0	823	938	115
0	950	1008	58
0	1049	1186	137
0	878	1015	137
0	854	991	137
0	877	1014	137
0	820	955	135
0	899	1036	137
0	855	893	38
0	827	964	137
0	925	960	35
0	900	1037	137
0	935	948	13
0	1005	1053	48
0	855	992	137
0	920	992	72
0	810	895	85
0	792	857	65
0	882	1019	137
0	934	1071	137
0	910	1047	137
0	865	973	110
0	899	1036	137
0	982	1119	137
0	856	993	137
0	961	963	2
0	893	1030	137
0	861	998	137
0	829	932	103
0	882	1019	137
0	875	1012	137
0	833	970	137
0	972	1013	41
0	807	944	137
0	924	959	35
0	845	982	137
0	840	977	137
0	867	1004	137
0	881	913	32
0	901	917	16
0	944	947	3
0	821	824	3
0	811	898	87
0	1007	1014	7

LOSS (O) OR DEATH (1)	AGE IN MO. AT ENTRY	AGE IN MO. AT EXIT	MO. SPENT IN STUDY
0	912	1049	137
0	802	939	137
0	928	1065	137
0	911	938	27
0	847	899	52
0	1035	1172	137
0	893	973	80
0	922	971	49
0	977	985	8
0	941	944	3
0	869	1006	137
0	885	955	70
0	859	996	137
0	948	1085	137
0	890	1005	115
0	887	891	4
0	968	1105	137
0	927	989	62
0	997	1134	137
0	846	983	137
0	831	861	30
0	842	979	137
0	768	905	137
0	896	1033	137
0	894	1014	120
0	885	1022	137
0	822	959	137
0	927	954	27
0	897	904	7
0	848	985	137
0	912	1001	89
0	863	1000	137
0	813	950	137
0	802	939	137
0	956	1061	105
0	822	945	123
0	934	993	59
0	1026	1054	28
0	981	1114	133
0	934	1071	137
0	836	927	91
0	760	897	137
0	820	957	137
0	907	1044	137
0	979	1016	37
0	894	1031	137
0	852	989	137
0	948	971	23
0	813	950	137
0	902	1024	122

LOSS (0) OR DEATH (1)	AGE IN MO. AT ENTRY	AGE IN MO. AT EXIT	MO. SPENT IN STUDY
0	913	1050	137
0	810	812	2
0	841	978	137
0	875	1012	137
0	841	927	86
0	948	1012	64
1	859	995	136
0	820	957	137
0	860	997	137
0	917	948	31
0	936	1073	137
0	950	986	36
0	1013	1031	18
0	847	984	137
0	777	914	137
0	960	988	28
0	920	1057	137
0	935	1051	116
0	933	979	46
0	933	985	52
0	797	934	137
0	733	870	137
0	866	953	87
0	870	930	60
0	795	875	80
1	905	1005	100
0	796	891	95
0	965	1102	137
0	775	912	137
0	942	977	35
0	895	926	31
0	981	1038	57
0	991	1006	15
0	894	918	24
0	886	943	57
0	871	924	53
0	839	976	137
0	839	976	137
0	858	995	137
0	830	967	137
0	868	1005	137
0	831	925	94
0	783	888	105
0	925	1062	137
0	898	1035	137
0	910	1009	99
0	958	1008	50
0	866	1003	137
0	851	988	137
0	906	1043	137

LOSS (0) OR DEATH (1)	AGE IN MO. AT ENTRY	AGE IN MO. AT EXIT	MO. SPENT IN STUDY
0	882	1019	137
0	815	952	137
1	972	1083	111
0	973	985	12
0	957	957	0
0	1010	1147	137
0	1070	1207	137
0	895	1032	137
0	818	860	42
0	864	1001	137
0	857	994	137
0	1028	1063	35
0	892	1029	137
0	769	906	137
0	883	1020	137
0	972	1109	137
0	965	1088	123
0	925	961	36
0	814	872	58
0	805	942	137
0	992	1010	18
0	943	1080	137
0	951	958	7
0	926	987	61
0	954	962	8
0	944	944	0
0	935	935	0
0	900	912	12
0	762	854	92
0	823	882	59
1	978	1010	32
1	966	1027	61
0	912	916	4
0	823	829	6
0	909	933	24
1	967	1041	74
1	851	905	54
1	843	861	18
1	963	1021	58
1	888	919	31
0	794	798	4
1	905	928	23
1	1039	1086	47
1	901	923	22
1	823	830	7
1	809	822	13
0	887	905	18
1	859	926	67
0	1004	1015	11
1	919	931	12

LOSS (0) OR DEATH (1)	AGE IN MO. AT ENTRY	AGE IN MO. AT EXIT	MO. SPENT IN STUDY
1	958	1041	83
1	1003	1093	90
1	871	944	73
1	864	873	9
1	996	1073	77
1	1034	1068	34
1	873	897	24
1	984	1047	63
1	943	1011	68
1	1007	1019	12
1	935	1006	71
1	929	996	67
1	939	978	39
1	772	840	68
0	871	912	41
1	873	954	81
1	981	1018	37
0	894	927	33
1	994	1040	46
1	976	995	19
1	847	883	36
1	859	941	82
1	933	990	57
1	861	934	73
1	886	908	22
1	943	986	43
1	931	969	38
1	948	1019	71
1	955	992	37
1	1004	1023	19
1	828	895	67
1	835	845	10
0	868	870	2
1	988	1074	86
1	861	930	69
0	959	976	17
1	859	912	53
0	871	874	3
0	847	892	45
1	874	885	11
1	992	1044	52
1	1027	1072	45
1	857	901	44
1	994	1013	19
1	1041	1043	2
0	900	926	26
1	1096	1152	56
1	930	936	6
1	943	994	51
1	1024	1063	39

```
                    APPENDIX 2:  CHANNING HOUSE WOMEN (CONT.)
  LOSS (0) OR       AGE IN MO.        AGE IN MO.      MO. SPENT
  DEATH (1)         AT ENTRY          AT EXIT         IN STUDY
      0               802               821              19
      0               811               819               8
      1               927              1001              74
      1               967               975               8
      1               943               982              39
      0               840               905              65
      1               979              1040              61
      0               921               926               5
      1               986              1030              44
      1              1039              1132              93
      1               968               990              22
      1               955               990              35
      1               837               911              74
      1               861               915              54
      1               967               983              16
```

EXPLORING THE INFLUENCE OF SEVERAL FACTORS ON A SET OF CENSORED DATA

SUE LEURGANS

<u>Medical Problem</u>. The influence of medical history in intrauterine device success rates.

<u>Investigator</u>. Louis Gordon, Alza Corporation, Palo Alto.

<u>Statistical Procedures</u>. Kaplan-Meier survival curves; Cox's censored data likelihood function.

MEDICAL BACKGROUND

Alza Corporation has been testing a new contraceptive device, called the Uterine Progesterone System (UPS). It is an intrauterine device (IUD) that releases progesterone in the uterus. Fewer side effects are anticipated from local release of progesterone than from release of the hormone into the bloodstream, as with pills. Alza ran a trial to test the contraceptive effectiveness of the device. They wish to use this data to identify characteristics of women who are prone to difficulties with the UPS.

Difficulty with the device can be described more precisely. Clearly, an accidental pregnancy is a failure of the device. There are other problems that can arise. Several of these are associated with the IUD nature of the device. Although the progesterone is contained in a small (and by itself, relatively ineffective) IUD, the problems common to all IUDs do occur. IUDs are sometimes spontaneously expelled or displaced. Some women experience increased pain and/or bleeding when wearing an IUD, and the device sometimes needs to be removed. For the purpose of

this study, these women are all regarded as having difficulty
with the device, and thus being women whom we would like to identify before they use the device.

THE DATA

4963 different women took part in the trial of the device.
The protocol of the experiment required that a medical history of
the patient be taken before insertion of the UPS. The information obtained emphasized contraceptive experience, pregnancies,
uterine characteristics, and menstrual patterns.

The general demographic information available is age and race.
Contraceptive history indicated whether the woman had had an IUD
in the last two years or had used the pill in the last three
months. Parity and elapsed time since the end of the last pregnancy (when applicable) constituted the pregnancy information.
Uterine size (length in centimeters) and angle were also
recorded. Premenstrual cramps, menstrual cramps, intermenstrual
cramps, and menstrual flow were all described on a scale from
none to severe. Two other variables were observed: whether or
not the woman was menstruating at insertion, and whether or not
she had previously had pelvic inflammatory disease (PID).[1]

Variables not directly related to the woman's physical condition are ignored in this analysis. These variables include identification numbers, lot codes, and physician codes. While it has
been observed that patients of different doctors have different
success rates, this is being ignored.

All of the above information was obtained before UPS insertion, and could therefore be used to screen prospective users.

The rest of the information can be used to determine whether
or not the woman had difficulty with the device. Each woman was
expected to return to her physician at specified times. Additional visits were scheduled as needed. Since the UPS was
designed to supply progesterone for one year, the physician was
to remove the UPS at the end of 12 months, replacing it with a
new device, if desired. (If a new device was inserted, this
insertion was ignored for the purposes of this work.) The dates
of all visits were recorded, as well as a code indicating what
transpired. This study used the information for the latest such
visit.

[1]To minimize the risk of fitting a pregnant woman with a UPS,
insertion was always between the third and eighth days of the
menstrual cycle. Some physicians believe that wearing an IUD may
make a woman more susceptible to PID.

THE TERMINATION CODES

The codes, known as termination codes, are essentially those recommended in Tietze and Lewin (1973) for use in studies of intrauterine contraception. The number in parentheses after each code is the number of women (out of 4963) to whom this code was assigned.

Accidental Pregnancy (69) - Code 1

If a UPS wearer became pregnant, the physician estimated the date of conception. If this date is after insertion and before either the physician removed the UPS or the wearer noticed expulsion or displacement, this termination code applies. This code also applies to all pregnancies in women with unnoticed displacement of the UPS and in women with a perforated uterus.

Expulsion (174) - Code 2

This code includes all displacements and expulsions, except those occurring during pregnancy or associated with conception.

Removal For Pain Or Bleeding (410) - Code 3

The UPS was removed because of pain and/or bleeding attributed by the wearer or by the physician to the UPS.

Removal For Other Medical Reasons (111) - Code 4

The medical reasons covered by this code include the following:

1. The wearer attributed other medical problems to the UPS.
2. The wearer was placed under treatment for a medical condition. Even when the medical condition was unrelated, the protocol required the removal of the device.
3. The wearer became pregnant, and conception was believed to have occurred before insertion of the UPS.
4. The uterus was perforated, and no pregnancy was associated.

Removal For Planning Pregnancy (54) - Code 5

This code applies when the wearer requested removal and stated that pregnancy was desired. Some of the people who work with such data suspect that an announced desire for pregnancy may be related to dissatisfaction with the device. Either the difficulties influence a reassessment of attitudes toward pregnancy, or women find this a comfortable way of telling the physician they want the device removed.

Removal For Other Personal Reasons (76) - Code 6

This code is used when the wearer requested removal for any reason not covered above. Thus this code includes those situations in which no reason was stated.

Release From Study (28) - Code 7

Typically, subjects released from the study were women who were moving and would be unable to keep appointments with their physician.

Documented Lost To Follow Up (94) - Code 8

At the last visit, the device was in place. The woman had missed an appointment and could not be contacted. (The protocol specified a procedure to use to try to contact the woman before assigning this code.)

Continuing At Last Visit (2330) - Code 9

At the last visit, the device was in place, and the woman was not documented lost to follow up. This code is not from Tietze and Lewin, and none of the women will have this code when the study is completed. However, since the data used for this work only included events through December, 1974, and since some insertions were made in early 1975, it is important at this stage.

Scheduled Removal (1597) - Code 10

The UPS was removed because it had been in place for a year, as intended. This code is not from Tietze and Lewin.
Some subsets of the data are given in Appendixes A and B.

STATISTICAL FORMULATION

Since Alza is primarily interested in identifying the women who have difficulty (rather than in exploring the causes of particular kinds of problems), it seems appropriate to simplify the termination codes 1-10 into failure and nonfailure, where failure (coded 1) is accidental pregnancy, expulsion, or removal for medical or personal reasons, and nonfailure is everything else, that is, release from study, documented lost to follow up, continuing at last visit, or scheduled removal. Examining the list of nonfailure codes, one can observe that they all apply when the experimenters cease to observe the woman wearing a UPS. This suggests that the problem has just been formulated as a censored data problem.

In such a problem, each woman is viewed as having two independently determined times: a failure time (possibly infinite) and a censoring time. If the failure time is less than the censoring time, we observe the failure. If, however, the censoring time is the smaller, all that we know about the failure time is that it is greater than the censoring time. Thus if a woman is classified as continuing at last visit, and her last examination was 173 days after insertion, we only know that she used the device for 173 days without difficulty.

We want to know the probability that women would have failures, if no censoring occurred. In other words, we want to estimate the cumulative distribution function (CDF) of the time until failure. More to the point, we want to know how this function varies with medical history.

The first approach used here involved grouping the women by a particular medical history variable and computing the Kaplan-Meier (1958) estimate of the CDF separately for each group. These curves were then compared heuristically. This approach was used to search for "important" variables.

The second approach consisted of an attempt to parametrize the hazard rate as in Cox (1972). This attempt foundered in computational seas.

THE KAPLAN-MEIER ESTIMATES

As described above, for each variable in turn the women were grouped according to their values of the variable under consideration. Then the Kaplan-Meier estimates of the CDF were computed (in the equivalent form of the survival function). Selected curves were then graphed (Figure 1), and the functions were tabled at representative time points. The results of the analyses for each variable are discussed in this section.

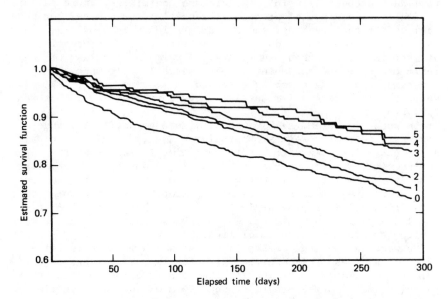

Figure 1.   The Kaplan-Meier curves for each parity.

Parity

Parity was the first variable to which this procedure was applied. It is relevant to note that in analyses of contraception data, it is standard practice to treat parous and nulliparous women separately. Thus it is not surprising that the curves obtained were different, and that increasing parity accompanied higher survival curves. In Table 1, as in all subsequent tables of Kaplan-Meier estimates, below each value is its estimated standard deviation.

Uterine Size

The Kaplan-Meier estimates seem to show that women with larger uteruses tend to be more successful (see Tables 2 and 3). It should be noted that women with uterine size less than 5 centimeters were not allowed to participate in the trial.

There is a considerable relationship between parity and uterine size. Since parity and uterine size are not independent, part of the differences of survival functions with parity are due to the influence of uterine size. From Table 4 we see that these

TABLE 1.   SELECTED VALUES OF KAPLAN-MEIER ESTIMATES OF PARITY
SURVIVAL FUNCTIONS

Time (days)	Parity					
	0	1	2	3	4	5
19	.9518	.9745	.9836	.9720	.9843	.9803
	.0066	.0049	.0036	.0065	.0078	.0074
94	.8650	.9092	.9188	.9256	.9492	.9481
	.0109	.0093	.0082	.0109	.0143	.0119
189	.7977	.8268	.8513	.8642	.9126	.8933
	.0132	.0129	.0112	.0153	.0194	.0178
275	.7429	.7641	.7810	.8318	.8399	.8528
	.0148	.0153	.0140	.0178	.0295	.0236
364	.6836	.7180	.7355	.8179	.8212	.8277
	.0163	.0169	.0156	.0189	.0316	.0270
Number of women	1098	1126	1353	718	280	385

variables are indeed related. However, from Tables 2 and 3, we
see that holding uterine size constant, the parous survival function is higher than the nulliparous survival function. Thus
while part of the influence of parity can be ascribed to differences in the distribution of uterine sizes, it appears that parity has an additional effect. No attempt was made to quantify
this relationship.

TABLE 2. SELECTED VALUES OF KAPLAN-MEIER ESTIMATES OF UTERINE
SIZE SURVIVAL FUNCTIONS FOR NULLIPAROUS WOMEN

Time (days)	Uterine Size (cm)					
	5-5.9	6-6.9	7-7.9	8-8.9	9-9.9	> 10
19	.8653	.9432	.9491	.9811	1.0	1.0
	.051	.013	.010	.009	–	–
94	.7681	.8515	.8657	.8922	.9231	1.0
	.065	.021	.016	.022	.074	–
189	.7150	.7740	.7956	.8487	.9231	.500
	.070	.025	.020	.027	.074	.354
275	.6576	.7230	.7235	.8282	.8392	.500
	.075	.028	.023	.029	.104	.354
364	.6576	.6625	.6525	.7914	.7552	.500
	.075	.030	.026	.032	.123	.354
Number of women	48	313	482	230	17	2

Age

From Tables 5 and 6 we see that this variable does not seem to
exhibit an effect that is constant across parity. Also, parity
and age are necessarily positively correlated. No attempt was
made to separate their influences.

Currently Menstruating

This variable does not, by itself, have any influence. The
values in Table 7 are typical. Even the standard errors are
close, reflecting the similarity of the curves and the group
sizes.

TABLE 3.   SELECTED VALUES OF KAPLAN-MEIER ESTIMATES OF UTERINE
SIZE SURVIVAL FUNCTIONS FOR PAROUS WOMEN

Time (days)	Uterine Size (cm)					
	5-5.9	6-6.9	7-7.9	8-8.9	9-9.9	> 10
19	.9831	.9701	.9792	.9811	.9705	1.00
	.017	.008	.003	.004	.011	-
94	.8116	.9009	.9196	.9483	.9052	.8300
	.051	.015	.007	.007	.020	.068
189	.6878	.8625	.8476	.8788	.8538	.7696
	.061	.018	.010	.011	.024	.072
275	.6046	.7979	.7880	.8264	.7761	.7696
	.067	.023	.012	.014	.030	.072
364	.5598	.7399	.7529	.8008	.7114	.7311
	.068	.027	.013	.015	.034	.078
Number of women	63	436	2014	1022	258	40

Variables Difficult to Study From This Data

This data set does not justify analyses of several variables
that did not vary much among the women. The binary variable
dealing with pelvic inflammatory disease does not give useful
curves, partly because only 48 nulliparous and 155 parous women
had had PID before entering the study. Table 8 displays non-
binary variables with substantially more than half the women in
one category. This problem is magnified by the presence of other
variables that have already been identified as relevant.

In each cell, the upper number is the appropriate number of
women. The lower number is the proportion of women of the same
parity group who gave the corresponding response.

TABLE 4.   UTERINE SIZE AND PARITY

| Uterine Size | Nulliparous | | Parous | | |
	Number of Women	Proportion	Number of Women	Proportion	Total
5-5.9	48	.044	63	.017	111
6-6.9	313	.287	436	.114	749
7-7.9	482	.441	2014	.528	2496
8-8.9	230	.211	1022	.268	1252
9-9.9	17	.016	258	.068	275
10	2	.002	40	.014	42
Total	1092		3813		4905

χ^2 Statistic = 265.3

We digress to describe how an apparent anomaly in the menstrual flow data was largely resolved.

There were a total of 22 women who reported no menstrual flow. This seemed rather odd, especially when the data for these women were examined. In particular, nine of these women were menstruating when the UPS was inserted. All the women were under 40. This seems quite contradictory.

The first suggested explanation was that many women interpreted the question as relating to the recent past. It was conjectured, therefore, that some women who had been pregnant would reply that they had no menstrual flow. Indeed, for 14 of the 19 parous women, a pregnancy had ended within 27 weeks. For only two of the parous women had more than a year passed since the end of their last pregnancy. Then it was noticed that both of these women had been using pills within the last three months. It is not unusual for menstruation to cease temporarily when a woman goes off the pill. Thus for the 19 parous women, no menstrual flow paradoxes remain.

There are three nulliparous women left. One of them was menstruating at insertion; none of them had been on pills within the last three months. Thus their flow observations cannot be

TABLE 5. SELECTED VALUES OF KAPLAN-MEIER ESTIMATES OF AGE
SURVIVAL FUNCTIONS FOR NULLIPAROUS WOMEN

Time (days)	Age (years)					
	≤ 19	20–25	26–30	31–35	36–50	> 40
19	1.00	.9402	.9538	.9675	.9400	1.00
	–	.016	.009	.011	.036	–
94	.7897	.8672	.8635	.8981	.8762	1.00
	.108	.023	.016	.019	.047	–
189	.6462	.7827	.8135	.8385	.7996	1.00
	.127	.028	.018	.023	.068	–
275	.6462	.7423	.7493	.8101	.6996	1.00
	.128	.031	.021	.025	.076	–
364	.6462	.6838	.6834	.7757	.6628	1.00
	.128	.033	.023	.027	.080	–
Number of women	15	247	549	289	54	6

explained from the data available. One can only suspect that
either these women had stopped using pills slightly more than
three months before insertion or that their flow variables were
not recorded accurately.

Other Variables

Menstrual cramps, TSELP, and contraceptive history variables
seem as if they could only be modeled by interaction terms or
nonlinear terms. For example, roughly speaking, women who used
either birth control pills within three months of insertion or an
IUD within two years of insertion did slightly better than the
women who used neither or both. However, few women used both.
With menstrual cramps, it appears that the extreme categories do

TABLE 6.   SELECTED VALUES OF KAPLAN-MEIER ESTIMATES OF AGE
SURVIVAL FUNCTIONS FOR PAROUS WOMEN

Time (days)	Age (years)					
	≤ 19	20-25	26-30	31-35	36-40	> 40
19	.9677	.9665	.9768	.9745	.9875	.9795
	.032	.012	.005	.005	.004	.008
94	.9021	.9146	.9099	.9157	.9429	.9265
	.054	.020	.010	.008	.008	.016
189	.7399	.7918	.8439	.8462	.8841	.8880
	.086	.031	.014	.012	.012	.020
275	.6936	.7507	.7872	.7812	.8232	.8362
	.093	.034	.017	.014	.015	.026
364	.6936	.7227	.7497	.7281	.7931	.8291
	.092	.037	.018	.016	.016	.027
Number of women	34	228	902	1292	1081	322

worse than those in the middle.   But the differences are not particularly convincing.   The curves for TSELP do not exhibit a
smooth relationship to that variable.

This technique is limited in that as more variables are considered simultaneously, the individual curves are based on smaller and often unevenly broken samples.   Useful results diminish
correspondingly.   And as more curves are compared, the possibility that spurious differences will be observed increases.

TABLE 7. SELECTED VALUES OF KAPLAN-MEIER CURVES FOR NULLIPAROUS
AND PAROUS WOMEN, MENSTRUATING AT INSERTION AND NOT MENSTRUATING
AT INSERTION

Time (days)	Nulliparous		Parous	
	Yes	No	Yes	No
19	.9536	.9501	.9804	.9756
	.009	.009	.003	.004
94	.8663	.8635	.9232	.9219
	.016	.015	.006	.007
189	.7973	.7978	.8555	.8545
	.019	.018	.009	.010
275	.7489	.7372	.8081	.7795
	.021	.021	.010	.013
364	.6956	.6726	.7793	.7294
	.038	.062	.032	.018
Number of women	519	577	2277	1574

PARAMETRIZATION

One way to avoid the small group sizes that limit the reduced
sample approaches is to adopt a parametric model. Therefore, an
attempt was made to apply a form of a model suggested by Cox
(1972). Since the model is easiest to understand when given in
terms of the hazard rate, we first explicitly transform the survival function problem into a hazard function problem.

The hazard function h(t) is defined as the ratio of the density f(t) to the survival function $\overline{F}(t) = 1 - F(t)$:

TABLE 8. CLUMPED VARIABLES

| | Response Categories | | | | | | | |
| | Nulliparous | | | | Parous | | | |
Variable	1	2	3	4	1	2	3	4
Premenstrual cramps	751	267	62	17	3031	610	168	43
	.685	.243	.056	.015	.787	.158	.043	.011
Race	1014	28	42	12	2501	783	335	183
	.925	.026	.038	.011	.658	.206	.088	.048
Intermenstrual cramps	1007	75	12	3	3610	183	49	7
	.918	.068	.011	.003	.938	.048	.013	.002
Flow	3	279	748	67	19	846	2739	242
	.003	.254	.682	.061	.005	.220	.712	.063
Uterine position	794	154	59		2283	823	808	
	.788	.153	.059		.583	.210	.206	

60

$$h(t) = \frac{f(t)}{\overline{F}(t)} = - \frac{d}{dt} \ln \overline{F}(t) \ . \tag{1}$$

Integrating both sides (assuming the necessary continuity conditions) and changing signs,

$$- \int_a^b h(t) \ dt = \int_a^b d \ln \overline{F}(t) = \ln \overline{F}(b) - \ln \overline{F}(a) \ . \tag{2}$$

Since we will be dealing with positive random variables, $\overline{F}(0) = 1$, and (2) can be specialized to

$$- \int_0^b h(t) \ dt = \ln \overline{F}(b) \tag{3}$$

or

$$\overline{F}(b) = \exp\left[-\int_0^b h(t) \ dt\right] \ . \tag{4}$$

Now suppose that different women have different hazard functions, and that these hazard functions are given by

$$h_i(t) = \theta(\underset{\sim}{z}_i) h_0(t) \tag{5}$$

where $\underset{\sim}{z}_i$ is the i-th woman's vector of medical history variables. Substituting in (4),

$$\overline{F}_i(t) = \exp\left[-\int_0^t \theta(\underset{\sim}{z}_i) h_0(t) \ dt\right]$$

$$= \exp\left[-\int_0^t h_0(t) \ dt\right]^{\theta(\underset{\sim}{z}_i)} = [F_0(t)]^{\theta(\underset{\sim}{z}_i)} \ . \tag{6}$$

The Cox model gives $\theta(\ \cdot\ )$ an exponential structure, postulating that

$$\theta(\underset{\sim}{z}_i) = e^{\underset{\sim}{\beta}' \underset{\sim}{z}_i} \ . \tag{7}$$

This gives

$$h_i(t) = e^{\underset{\sim}{\beta}'\underset{\sim}{z}_i} h_0(t) \ . \tag{8}$$

The program available when the first fits were made was designed for time-dependent covariates. It computed many expressions of the form

$$T(i) = \sum_{j \in \mathcal{R}_i} \exp[\underset{\sim}{b}'\underset{\sim}{z}_j(t_i)] \tag{9}$$

where $\mathcal{R}_i$ = risk set at time i = {all women still under observation at time t_i}. When the time dependence in (9) is eliminated, as in this problem, we have

$$\tilde{T}(i) = \sum_{j \in \mathcal{R}_i} \exp(\underset{\sim}{b}'\underset{\sim}{z}_j) \ . \tag{10}$$

Here, there is a clear recursive relationship:

$$\tilde{T}(i+1) = \frac{\tilde{T}(i)}{\sum\limits_{j \in \mathcal{R}_i / \mathcal{R}_{i+1}} \exp(\underset{\sim}{b}'\underset{\sim}{z}_j)} \ . \tag{11}$$

Thus the $\tilde{T}$'s can be computed more rapidly than the T's. This implies that the non-time-dependent computations are theoretically easier than the time-dependent computations. Unfortunately, a reliable non-time-dependent program was not available while this work was being done. It was ascertained that the time-dependent version is much too slow to be appropriate for a data set of this size.

In conclusion, one set of results that was obtained is discussed.

A Cox model was fitted for the nulliparous women, with uterine size measurement recoded, assigning uterine size measurements less than 6 centimeters to -1, those between 6 and 7.9 centimeters to 0, and those between 8 and 8.9 centimeters to +1. Missing measurements were coded 1/6, which was the mean of the coded measurements.

After three iterations, the program terminated because the change in the estimate of the parameter was a sufficiently small proportion of the estimated standard deviation. β was estimated to be -.4319, with estimated standard deviation .129. Since the fitted model estimates β to be negative, we see from (6) that the

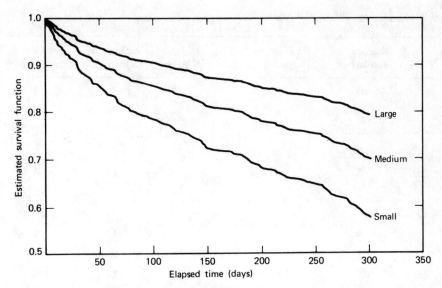

Figure 2.   Estimated survival function obtained from fitting Cox model.   Covariate: uterine size.   (Nulliparous women.)

fitted model predicts that women with larger uterine sizes will be more successful.   These curves are plotted in Figure 2.

REFERENCES

Cox, D. R. (1972). Regression models and life-tables. J. Roy. Stat. Soc., Ser. B, 62, 187–202.

Kaplan, E. L. and P. Meier (1958). Nonparametric estimation from incomplete observations. J. Amer. Statist. Assoc., 53, 457–481.

Tietze, C. and S. Lewin (1973). Recommended procedures for the statistical evaluation of intrauterine contraception. Stud. Fam. Plan., 4, 35–42.

APPENDIX A: NULLIPAROUS WOMEN

Time	Number of Failures	Number of Censors	Time	Number of Failures	Number of Censors
1	10	59	35	2	4
2	3	9	36	2	3
3	1		37	1	1
4	3		38	1	4
5	3		40	1	4
6	2		41	1	2
7	3		43	6	10
8	4		47	1	1
9	2		50	5	5
10	1		51	1	4
11	5	1	52	1	4
12	2		54	2	
14	1		55	2	
15	4	1	56	2	1
17	2		57	1	10
18	4		63	2	
21	2		64	2	2
22	4		65	4	1
23	1		67	3	1
24	1	1	68	1	
26	1	1	71	2	1
27	2		73	1	5
28	1	3	76	2	
29	9	21	77	1	2
30	3	4	78	3	6
31	1	4	82	1	1
33	1	4	83	1	1

APPENDIX A (Continued)

Time	Number of Failures	Number of Censors	Time	Number of Failures	Number of Censors
84	1	6	145	2	
88	1	2	146	1	
91	2	5	147	2	
94	1	1	148	3	2
95	1		154	1	
96	1	1	155	1	2
97	1	4	159	1	14
100	1	2	172	2	
102	1	1	173	2	2
105	2	4	176	1	8
107	2		182	1	1
109	1		183	2	6
113	1	4	184	1	
115	1		185	1	3
116	1	3	186	1	1
117	1	1	188	2	
120	4	4	190	3	7
127	2	1	193	1	2
128	1		196	1	3
129	1		198	1	
133	1		199	2	3
134	2	1	201	1	11
135	1		211	2	4
136	1	1	213	1	
141	3	1	214	2	1
142	1	1	217	1	
144	1		218	1	3

65

APPENDIX A (Continued)

Time	Number of Failures	Number of Censors	Time	Number of Failures	Number of Censors
219	1	3	289	1	1
221	1		291	2	
225	3	6	294	3	
235	1	3	295	2	
238	1		296	1	
240	1	1	298	1	
241	1	5	301	2	
245	1	3	302	1	
250	1	1	303	1	7
256	1	3	310	2	
257	3	3	312	2	2
260	2	1	316	1	3
262	2		323	1	5
263	2		329	1	
264	1	1	330	1	5
265	1	1	331	2	1
267	1	1	332	2	1
270	1		333	1	2
276	1	1	336	1	2
277	1		337	1	102
279	1		361	1	219
281	3	1	389	1	17
282	2	1	395	1	80
286	1	2	442	3	25
288	2		524	1	9

APPENDIX B: PAROUS WOMEN

Time	Number of Failures	Number of Censors	Time	Number of Failures	Number of Censors
1	2	357	30	4	14
2	6	21	31	5	21
3	3		32	9	23
4	5		33	5	17
5	3		34	2	13
6	2		35	7	17
7	7	1	36	9	43
8	4	1	37	1	16
9	4		38	4	17
10	3		39	2	12
11	6		40	4	12
12	3	1	41	3	11
13	4		42	1	6
14	4		43	8	26
15	7		44	3	10
16	4		45	1	8
17	2		46	1	7
18	4		47	4	3
19	2		48	1	3
21	2		49	1	10
22	3	1	50	4	17
23	2	5	52	2	12
25	2	3	54	1	6
26	3	5	57	5	12
27	1	4	59	1	9
28	2	13	61	1	1
29	11	47	62	3	8

67

APPENDIX B (Continued)

Time	Number of Failures	Number of Censors	Time	Number of Failures	Number of Censors
64	6	3	99	1	10
65	1	1	100	2	4
66	1	1	101	1	2
67	3	10	102	1	8
68	1	1	103	1	12
69	3	2	105	1	6
70	3	3	106	3	9
71	3	12	107	2	3
72	2	4	108	3	8
73	1	1	111	6	1
74	2	3	112	3	1
76	4	20	113	8	8
78	3	1	114	2	2
79	2	1	115	1	2
80	3	8	116	2	3
83	3	9	117	1	5
85	2	6	119	2	5
86	1	7	120	2	15
88	2	6	123	1	12
89	2	4	125	1	3
90	2		126	2	3
91	3	6	127	3	5
92	5	17	128	3	2
93	1	10	129	1	
95	2	7	130	2	4
96	1	27	131	2	3
98	4	6	132	2	10

APPENDIX B (Continued)

Time	Number of Failures	Number of Censors	Time	Number of Failures	Number of Censors
134	5	15	166	3	6
135	1	5	168	2	7
136	1	4	169	5	11
137	1	6	170	1	1
138	2	6	171	1	2
139	7	3	172	3	13
140	1	4	174	1	7
141	2	3	175	1	9
143	1		176	10	12
144	2	1	177	1	3
145	3		178	2	8
146	1	6	179	4	17
148	2	6	182	3	16
149	1		183	3	30
150	2	4	184	1	22
152	1	1	185	5	13
153	1	2	186	6	24
155	4	1	187	1	13
156	1	4	188	2	10
158	1	4	189	3	12
159	2	4	190	5	25
160	1	2	191	1	9
161	2	6	192	1	7
162	3	10	193	1	5
163	1	4	194	1	9
164	1	10	196	1	41
165	2	7	199	1	4

APPENDIX B (Continued)

Time	Number of Failures	Number of Censors	Time	Number of Failures	Number of Censors
200	2	9	238	1	3
201	2	16	239	4	4
203	3	4	240	1	4
204	6	14	241	1	4
205	2	26	242	1	2
210	1	5	243	2	4
211	3	11	245	1	2
212	2	5	246	2	7
214	2	3	248	1	3
215	4	22	251	1	3
218	4	11	252	3	
219	1	6	253	3	10
220	2	13	256	1	2
221	4	10	257	1	2
222	1	6	259	1	3
223	1	13	260	2	5
225	1	10	261	1	10
226	3	7	265	1	1
228	1	9	267	1	5
229	2	3	268	4	
230	2	2	269	2	1
231	3	18	270	2	1
232	1	10	271	1	1
233	2	2	272	1	
234	1	6	273	1	4
235	4	10	275	1	
237	1		277	1	1

APPENDIX B (Continued)

Time	Number of Failures	Number of Censors		Time	Number of Failures	Number of Censors
278	2	1		323	1	4
279	2			325	1	1
280	2	1		326	1	4
281	1			327	1	1
282	1			329	2	3
284	4	2		331	1	16
288	2	2		337	1	57
289	2			349	1	72
290	1	2		355	1	139
293	1			363	1	16
294	2			364	2	16
295	1	2		365	2	463
297	3	4		380	1	13
301	1			381	1	68
302	4	3		388	1	16
303	2	2		390	1	102
305	1			401	1	7
306	1	1		402	1	19
308	2	1		404	1	17
310	1	1		407	1	37
312	2			411	1	30
313	2	2		415	1	24
314	1	1		420	1	26
316	3	1		428	1	41
317	3	1		443	1	31
319	1	4		463	1	15
322	2			471	1	32

APPENDIX B (Continued)

Time	Number of Failures	Number of Censors	Time	Number of Failures	Number of Censors
523	1	17	623	1	

COMBINING 2 × 2 CONTINGENCY TABLES

RUPERT G. MILLER, JR.

Medical Problem. A possible relationship between tonsillectomy and infectious mononucleosis.

Medical Investigators. Richard Goode and Donald Coursey, Stanford University.

Statistical Procedures. Combining 2 × 2 contingency tables; relative risk analysis; Mantel-Haenszel test; McNemar's test.

MEDICAL BACKGROUND

Recent research has indicated that a virus, the Epstein-Barr virus (EBV), causes infectious mononucleosis. Some investigators feel that the EBV resides and replicates in the oropharynx with transmission via the buccal fluids. The tonsillar or adenoid lymphoid tissue could possibly serve as a reservoir and replicating milieu for the EBV. Tonsillectomy and adenoidectomy might therefore eliminate this reservoir.

To check this hypothesis Goode and Coursey surveyed students seen at Stanford University's Cowell Student Health Center. They reviewed the charts of 203 students treated for infectious mononucleosis between January 1968 and May 1973 for definite confirmation of the disease and for determination of any history of tonsillectomy. This constitutes the IM group. The control group consists of 989 students seen at Cowell Health Center between April and September 1973 who came in for any ailment whatsoever and who were willing to note on a survey sheet whether or not they had had a tonsillectomy. It is possible that one or more of the IM group might have filled in the survey sheet and thereby be counted in both groups, but the number would be so small in comparison to the size of the control group as to make no difference.

The objective of Goode and coursey was to compare the percentage of students having undergone a tonsillectomy in the IM group with the control group percentage. Goode and Coursey knew how to compute the χ^2 statistic for a 2 × 2 contingency table, but they were worried about two associated problems so they came to a statistical consultant.

The first worry was that the ages of the students might be affecting the comparison. The control group contains a larger portion of old students and older students have had more opportunity to undergo a tonsillectomy. Also, medical fashion on whether or not to perform a tonsillectomy has varied over the years. In addition, it was conceivable, but not as likely a priori, that sex is another variable influencing the comparison.

The second concern to Goode and Coursey was a prior series of papers investigating the relationship between tonsillectomy and Hodgkin's disease, which had erupted into a statistical rhubarb. Vianna et al. (1971) published a paper showing that a group of Hodgkin's patients had a significantly higher percentage of prior tonsillectomies than did a group of comparable controls. They speculated that the tonsils act as a filter and therefore tonsillectomy would increase the risk of Hodgkin's disease by removal of the lymphoid barrier. Johnson and Johnson (1972) challenged this by exhibiting statistics on a series of patients of their own which they claimed showed no difference in the percentages between the Hodgkin's and control groups. The Johnson–Johnson paper was followed by three Letters to the Editor by Cole et al. (1973), Pike and Smith (1973), and Shimaoka et al. (1973), which pointed out an error in the Johnson–Johnson analysis and supported the Vianna et al. conclusion. Goode and Coursey naturally did not want to become embroiled in a statistical controversy so they checked with a statistician to insure that there was no fallacy in their analysis.

Both the tonsillectomy versus infectious mononucleosis data and the tonsillectomy versus Hodgkin's disease data are discussed in this report. The former is illustrative of relative risk analysis and the latter is illustrative of the proper analysis for 2 × 2 contingency tables with matched subjects.

DATA AND STATISTICAL ANALYSIS

Tonsillectomy Versus Infectious Mononucleosis

Goode and Coursey compiled the following 2 × 2 contingency table:

	T	No T	
IM	44	159	203
C	409	580	989
	453	739	1192

(1)

(IM = infectious mononucleosis, C = control, and T = tonsillectomy.) The control group contained somewhat older students so to equalize the age effect Goode and Coursey compiled the corresponding table for 18 to 24 year olds:

	T	No T	
IM	40	145	185
C	235	420	655
	275	565	840

(18-24 years)   (2)

In (1) the percentage of IM with prior tonsillectomies is 22% and for the controls it is 41%. The χ^2 value corrected for continuity with 1 df is 26.86, which is highly significant. For the age-restricted groups the percentages are 22 and 36%. The difference between the percentages is less, but the χ^2 value of 12.70 is still significant at the 1% level. This seems to confirm the hypothesis that tonsillectomy reduces the risk of infectious mononucleosis.

To investigate the possible sex and age effects in greater detail several additional analyses were performed. Students outside the age range 18-24 were ignored because so very few of the IM group fell outside this range that comparison between the two groups for other ages was impossible. Within the 18-24 range the students were divided into 14 groups by year of age and by sex. Eyeball comparison of differences in the percentages due to sex revealed nothing with the exception that 19-year-old females in the control group had a low percentage of tonsillectomies. It

was decided this was a random artifact, and the sexes were pooled
to give the following table of fractions of students with
tonsillectomies:

	18	19	20	21	22	23	24	
IM	6/23	3/42	12/41	8/46	5/15	2/9	4/9	(3)
C	17/49	26/96	34/112	48/139	45/118	29/66	36/75	

From (3) the percentage of students with tonsillectomies in each
age and patient group were as follows:

	18	19	20	21	22	23	24		
IM	26	07	29	17	33	22	44	%	(4)
C	35	27	30	35	38	44	48	%	

The significance of the differences in the percentages between
the IM and C groups in (4) was assessed by two different relative
risk analyses: the logit approach and the Mantel-Haenszel
approach (see Armitage, 1971, pp. 427-433).
   Consider both a general 2 × 2 table and specifically the 2 × 2
table for 18 year olds:

	T	No T	
IM	a	b	
C	c	d	(5)
		n	

	T	No T		
IM	6	17	23	
C	17	32	49	(6)
	23	49	72	

The estimated cross-product ratio or approximate relative risk is

$$\hat{\psi} = \frac{ad}{bc} = \frac{6 \times 32}{17 \times 17} = .66 \ . \tag{7}$$

The estimate .66 is the approximate decrease in risk of contracting infectious mononucleosis after tonsillectomy.

The logarithm of the approximate relative risk is the difference between the logits for the two groups, that is,

$$\ln \hat{\psi} = \ln \frac{ad}{bc} = \ln \frac{a}{b} - \ln \frac{c}{d} = \ln \frac{\hat{p}_1}{1-\hat{p}_1} - \ln \frac{\hat{p}_2}{1-\hat{p}_2} \, . \tag{8}$$

The asymptotic variance of (8) (see Cox, 1970, pp. 30-32) is estimated by

$$\hat{\mathrm{Var}}(\ln \hat{\psi}) = \frac{1}{a} + \frac{1}{b} + \frac{1}{c} + \frac{1}{d} \, . \tag{9}$$

Woolf (1955) proposed that the logarithms of the approximate relative risks (8) from different 2 × 2 tables can be combined by weighting them inversely proportional to their variances (9). For the seven 2 × 2 tables in (3) the estimated approximate relative risks, their logarithms, variances, and standard deviations are given below:

	18	19	20	21	22	23	24
$\hat{\psi}$	.66	.21	.95	.40	.81	.36	.87
$\ln \hat{\psi}$	-.42	-1.56	-.05	-.92	-.21	-1.02	-.14
$\hat{\mathrm{Var}}(\ln \hat{\psi})$	.32	.41	.16	.18	.34	.70	.50
$\hat{\mathrm{SD}}(\ln \hat{\psi})$	.57	.64	.40	.42	.58	.84	.71

(10)

The combined estimate is

$$\ln \hat{\psi} = \frac{\Sigma \ln \hat{\psi}_i \times \hat{V}_i^{-1}}{\Sigma \hat{V}_i^{-1}} = -.54 \, . \tag{11}$$

The approximate variance of (11) is

$$\hat{\mathrm{Var}}(\ln \hat{\psi}) = 1/\Sigma \, \hat{V}_i^{-1} = .042 = (.205)^2 \tag{12}$$

so the estimate (11) differs significantly (P ≅ .01) from the null value ln 1 = 0. Taking the antilogarithm of (11) gives an estimate of the approximate relative risk:

$$\hat{\psi} = \exp(\ln \hat{\psi}) = .58 \ . \tag{13}$$

This compares closely with the estimate $\hat{\psi} = .49$ obtained from the combined table (2).

The Mantel-Haenszel (1959) approach computes a different quantity for each 2 × 2 table and then combines the results. It compares each upper-left-hand cell frequency [i.e., a in (5)] with its expected value under the hypothesis of no difference between groups, which is

$$(a+b) \times \frac{(a+c)}{n} \ . \tag{14}$$

The variance of a under the null hypothesis is

$$\frac{(a+b)(c+d)(a+c)(b+d)}{n^2(n-1)} \ . \tag{15}$$

For the seven 2 × 2 tables in (3) these quantities are given below:

	18	19	20	21	22	23	24
a	6	3	12	8	5	2	4
E(a)	7.35	8.83	12.33	13.92	5.64	3.72	4.29
Var(a)	3.45	4.89	6.35	7.33	3.15	1.95	2.03
SD(a)	1.86	2.21	2.52	2.71	1.77	1.40	1.42

$$\tag{16}$$

The Mantel-Haenszel statistic for testing the difference between groups is

$$\frac{\left| \Sigma[a_i - E(a_i)] \right| - 1/2}{\sqrt{\Sigma \ \mathrm{Var}(a_i)}} = \frac{15.58}{5.40} = 2.89 \ . \tag{17}$$

Thus, the groups are significantly different with $P < .01$ from normal tables. As an estimate of the approximate relative risk Mantel and Haenszel propose

$$\frac{\Sigma \; a_i d_i / n_i}{\Sigma \; b_i c_i / n_i} = .55 \; , \tag{18}$$

which agrees well with the Woolf estimate $.58$ and the combined 2×2 table estimate $.49$.

On the basis of these analyses the conclusion has to be that tonsillectomy reduces the risk of contracting infectious mononucleosis by a factor estimated to be approximately one-half.

The Woolf and the Mantel-Haenszel methods are the two most common procedures for handling a series of 2×2 tables, but there are alternatives. To test the significance between groups the simple sign test gives a one-sided $P = (1/2)^7 = .008$ since the cross-product ratio is less than 1 in all seven cases. Individual one-sided P values could be computed for each of K 2×2 tables and the results combined through (1) Fisher's $\Sigma - 2 \ln P_i$ or (2) $\Sigma \; Z_i / \sqrt{K}$, where $Z_i = \Phi^{-1}(P_i)$, Φ = normal cdf. Fisher's test can be badly biased toward nonsignificance by the discreteness of the distribution (see Lancaster, 1949 and Pearson, 1950), but roughly, the first statistic is referred to a χ^2 table with 2K df, the second to a normal table. None of these methods, however, give a combined estimate of the relative risk like Woolf or Mantel-Haenszel. The problem could also be approached from the point of view of a $K \times 2 \times 2$ contingency table.

Tonsillectomy Versus Hodgkin's Disease

Vianna et al. (1971) considered a series of 109 Hodgkin's patients. By means of hospital records 109 control patients were selected to generally match the Hodgkin's group composition on the basis of age, sex, race, county of residence, date of hospital admission, and absence of malignant disease or chronic illness. For eight Hodgkin's patients and two control patients the tonsillectomy history could not be determined so these patients were eliminated from consideration. The 2×2 contingency table for the remaining 208 patients was the following:

$$
\begin{array}{c c}
 & \text{T} \quad \text{No T} \\
\text{H} & \boxed{\begin{array}{c|c} 67 & 34 \end{array}} \quad 101 \\
\text{C} & \boxed{\begin{array}{c|c} 43 & 64 \end{array}} \quad 107 \\
 & 110 \quad 98 \quad 208
\end{array}
\qquad (19)
$$

(H = Hodgkin's disease, C = control, and T = tonsillectomy.)

The χ^2 value corrected for continuity is 13.23, which is significant with P < .01. The estimate of the approximate relative risk (7) is

$$
\hat{\psi} = \frac{67 \times 64}{43 \times 34} = 2.93 . \qquad (20)
$$

Thus the conclusion was that tonsillectomy increases the chance of contracting Hodgkin's disease by a factor of nearly 3. The interpretation by Vianna et al. was that a protective barrier has been removed.

Johnson and Johnson (1972) examined a series of patients of their own. They reviewed 175 consecutive patients treated at the Radiation Branch of the National Cancer Institute for Hodgkin's disease. For one patient the tonsillectomy history was not available. As controls Johnson and Johnson used siblings of the Hodgkin's patients. There was information available on a total of 472 siblings at risk for 172 of the 174 patients. The authors chose the closest sibling of the same sex and within five years of the same age to act as a matched pair. This matching reduced the data to 85 patient-sibling pairs with the following breakdown on tonsillectomies:

$$
\begin{array}{c c}
 & \text{T} \quad \text{No T} \\
\text{H} & \boxed{\begin{array}{c|c} 41 & 44 \end{array}} \quad 85 \\
\text{C} & \boxed{\begin{array}{c|c} 33 & 52 \end{array}} \quad 85 \\
 & 74 \quad 96 \quad 170
\end{array}
\qquad (21)
$$

For table (21) the usual χ^2 statistic has the value 1.17 with a corresponding P value of .28. On the basis of this and other comparisons Johnson and Johnson claimed to have refuted the contention of Vianna et al. that the tonsils constitute a lymphoid barrier to Hodgkin's disease.

There is a basic error in the Johnson–Johnson statistical analysis, and this was pointed out in Letters to the Editor by Cole et al. (1973), Shimaoka et al. (1973), and Pike and Smith (1973). When matched pairs are involved as in (21), the usual probability structure assumed for 2 × 2 contingency tables is not satisfied and the usual χ^2 test is inappropriate.

All the information on the association between tonsillectomy and Hodgkin's disease resides in the pairs where the patient and sibling have different histories of tonsillectomy. The correct way of compiling the 2 × 2 table which takes account of the pairing is as follows:

Sibling

		T	No T	
Hodgkin's patient	T	26	15	41
	No T	7	37	44
		33	52	85

(22)

If there is no association, then the probability is 1/2 that a pair falls in the upper right cell (b') and 1/2 that it falls in the lower left cell (c'), given that the pair falls off the main diagonal. Since the pairs are independent, the ratio 15/22 can be compared with 1/2 by a binomial test. This test procedure is due to McNemar (see Siegel, 1956, pp. 63–67).

When the number of off-diagonal terms is large enough to use the normal approximation to the binomial, the computation is very simple:

$$\left(\frac{b'/(b'+c') - 1/2}{\sqrt{[1/(b'+c')] \times 1/2 \times 1/2}}\right)^2 = \frac{(b'-c')^2}{b'+c'} .$$

(23)

With a continuity correction this becomes

$$\chi^2 = \frac{(|b'-c'|-1)^2}{b'+c'} = \frac{(|15-7|-1)^2}{15+7} = 2.23 ,$$

(24)

which with 1 df has a P value of .14. Although this P value does not reach the 5% level of statistical significance, it is suggestively small and the estimate of the approximate relative risk,

$$\hat{\psi} = \frac{b'}{c'} = \frac{15}{7} = 2.14 \ , \tag{25}$$

is not in serious disagreement with the Vianna et al. estimate of 2.93.

If the individual pairings are ignored and all siblings are used collectively as a control group, the following 2 × 2 table emerges:

	T	No T	
H	90	84	174
C	165	307	472
	255	391	646

(26)

Although it is not valid to use the usual χ^2 test because of the pairing involved, it is interesting that for table (26) $\chi^2 = 14.27$ and the approximate relative risk estimate is $\hat{\psi} = 1.99$.

Rather than contradicting the Vianna et al. hypothesis, the Johnson and Johnson data tend to confirm it. Johnson and Johnson in essence have been hoisted with their own petard.

CONCLUSION

Goode and Coursey were assured that they had not fallen into the Johnson and Johnson statistical trap. They had not paired their subjects on influential predictor factors and then ignored the factors and pairing in their analysis. The Johnson and Johnson trap can easily lead to grief through incorrect accept-ance of the null hypothesis.

Goode and Coursey were justified in worrying about systematic differences in age and other factors between their unmatched study and control groups. Such differences could indeed cause rejection of the null hypothesis for the wrong reasons. However, standard methods of adjusting for age did not erase the differ-ence they had uncovered and their conclusion that tonsillectomy protects against infectious mononucleosis seems justified by their data.

REFERENCES

Armitage, P. (1971). Statistical Methods in Medical Research.
New York: Wiley.

Cole, P., T. Mack, K. Rothman, B. Henderson, and G. Newell
(1973). Tonsillectomy and Hodgkin's disease. New Engl. J.
Med., 288, 634.

Cox, D. R. (1970). The Analysis of Binary Data. London:
·Methuen.

Johnson, S. K. and R. E. Johnson (1972). Tonsillectomy history
in Hodgkin's disease. New Engl. J. Med., 287, 1122-1125.

Lancaster, H. O. (1949). The combination of probabilities aris-
ing from data in discrete distributions. Biometrika, 36,
370-382.

Mantel, N. and W. Haenszel (1959). Statistical aspects of the
analysis of data from retrospective studies of disease. J.
Nat. Cancer Inst., 22, 719-748.

Pearson, E. S. (1950). On questions raised by the combination of
tests based on discontinuous distributions. Biometrika, 37,
383-396.

Pike, M. C. and P. G. Smith (1973). Tonsillectomy and Hodgkin's
disease. Lancet, 1, 434.

Shimaoka, K., I. D. J. Bross, and J. Tidings (1973). Tonsillectomy and Hodgkin's disease. New Engl. J. Med., 288, 634-
635.

Siegel, S. (1956). Nonparametric Statistics for the Behavioral
Sciences. New York: McGraw-Hill.

Vianna, N. J., P. Greenwald, and J. N. P. Davies (1971). Tonsillectomy and Hodgkin's disease: the lymphoid tissue barrier.
Lancet, 1, 431-432.

Woolf, B. (1955). On estimating the relation between blood group
and disease. Ann. Hum. Genet., 19, 251-253.

RECOVERING INFORMATION FROM FRAGMENTARY DATA

LINCOLN E. MOSES AND
SUZANNA WONG

Problem. How are aptitude scores affected by short term instructive efforts, race, sex, and other factors?

Investigator. Franklin Evans, Educational Testing Service, Princeton.

Statistical Procedures. Confidence intervals; t tests; various descriptive statistics

BACKGROUND AND PROBLEM

Aptitude tests are intended to assess rather stable, slowly changing abilities, while achievement tests (e.g., course examinations) are aimed at assessing qualities that are presumably susceptible to short-term change through learning (or forgetting). But it is possible that coaching can improve scores on aptitude tests even if those tests are intended to measure "stable" qualities. The Graduate Record Examination Board research program conducted an experiment directed to just this point.

On 11 campuses instructors were recruited (17 of them in all) to give courses (meeting eight times at weekly intervals) in accordance with a certain experimental plan. Students of three races were offered free GRE examinations as an incentive for entering the study. To increase precision by taking account of variation in student abilities the first session was a pretest. After that the schedule differed for experimental and control subjects. The experimental subjects received a final test in the eighth week following their instruction, while the control subjects received the corresponding test <u>before</u> their instruction. An additional feature was that both sets of subjects were given, for their first instruction session, training that was not substantively germane to the test content, but rather to the testing situation. That instruction is here labeled "Anxiety Reduction Instruction." All subjects received another test (the "Anxiety Reduction Test") at the session next after the anxiety reduction instruction. The schedule was:

		Experimentals	Controls
	1	Pretest	Pretest
	2	Anxiety reduction instruction	Post-test
Week	3	Anxiety reduction test	Anxiety reduction instruction
	4	Coaching	Anxiety reduction test
	5-7	Coaching	Coaching
	8	Post-test	Coaching

Every subject, thus, had three sessions of testing, one session of anxiety reduction instruction, and four sessions of coaching in the kinds of content assessed by the tests. The scores of the experimental subjects could be used to assess the effects of coaching, based both on comparison with pretest scores and with scores obtained after the anxiety reduction phase. The scores of the control subjects could be used to assess the effects of anxiety reduction instruction.

It was arranged that each instructor would teach both an experimental and a control class, and that the students who volunteered for the instruction would be randomly assigned to one of the two classes. Classes were expected to have up to 30 subjects, ideally with students of both sexes and with at least five from each of three racial groups. The experiment was planned to be large; 17 instructors each with two classes of 30 subjects would have involved more than 1000 students.

Unfortunately, failure to fill classes and dropouts after instruction began reduced the experiment most cruelly. In all, 446 students entered the study. Of these 113 dropped out after

the pretest and only 145 finished complete records. A summary
account appears in Table 1, where for each race the numbers of
subjects completing various combinations of the three tests are
shown.

A second summary, that in Table 2, is more revealing in some
respects. There are two rows for each instructor, one for the
experimental class, one for the control class. (Vertical tie
marks on the left indicate pairs of instructors who are in the
same college.) The first three columns correspond to race, and
the fourth column gives totals. The entries in the table show
how many students there were who took at least one of the three
tests, together with the number (in parentheses) who took all
three tests, that is, the number of complete cases. The degree
to which the data are fragmentary shows clearly in this table!
Instructors 7 and 10 were alone in having each racial group
represented by one or more complete case--and neither of these
instructors had a complete case for each racial group in both the
experimental and control classes.

The statistical problem to be addressed is: What conclusions,
if any, can be drawn about the effects of coaching and of the
anxiety reduction instruction? Additional interpretations with
regard to race and sex are desired if possible.

TABLE 1.   SUMMARY OF AVAILABLE DATA SHOWING NUMBERS OF SUBJECTS
BY RACE, COMPLETING VARIOUS TESTS

Tests Completed	White	Black	Chicano	All
Pre, Anxiety reduction, Post	85	46	14	145
Pre, Anxiety reduction	26	23	7	56
Pre, Post	58	30	29	117
Anxiety reduction, Post	3	2	0	5
Pre	41	52	20	113
Anxiety reduction	4	3	1	8
Post	1	0	1	2
Totals	218	156	72	446

TABLE 2. DISTRIBUTION OF SUBJECTS (BY RACE) FOR THE 17
INSTRUCTORS WITH NUMBERS OF COMPLETE CASES SHOWN IN PARENTHESES

Instructor Number[a]	Class	White	Black	Chicano	All
1	Experimental	0	18(4)	0	18(4)
	Control	0	18(5)	0	18(5)
2	Experimental	11(0)	0	3(0)	14(0)
:	Control	16(0)	0	3(0)	19(0)
3	Experimental	9(0)	1(0)	3(0)	13(0)
	Control	10(0)	0	1(0)	11(0)
4	Experimental	6(2)	6(5)	1(0)	13(7)
:	Control	9(5)	0	0	9(5)
5	Experimental	6(3)	1(0)	0	7(3)
	Control	4(2)	4(2)	0	8(4)
6	Experimental	3(0)	0	11(0)	14(0)
	Control	3(0)	0	7(0)	10(0)
7	Experimental	9(5)	0	11(7)	20(12)
:	Control	15(8)	1(1)	10(6)	26(15)
8	Experimental	11(0)	0	7(0)	18(0)
	Control	5(0)	0	9(0)	14(0)
9	Experimental	11(3)	4(0)	1(0)	16(3)
:	Control	8(5)	3(2)	0	11(7)
10	Experimental	9(4)	9(1)	1(1)	19(6)
	Control	5(1)	8(1)	0	13(2)
11	Experimental	17(7)	0	0	17(7)
	Control	19(16)	0	0	19(16)
12	Experimental	0	17(6)	0	17(6)
	Control	0	19(11)	0	19(11)
13	Experimental	4(3)	3(2)	0	7(5)
	Control	5(3)	2(1)	0	7(4)

TABLE 2 (Continued)

Instructor Number[a]	Class	White	Black	Chicano	All
14	Experimental	0	14(0)	2(0)	16(0)
⋮	Control	0	13(2)	2(0)	15(2)
15	Experimental	0	0	0	0
	Control	0	1(0)	0	1(0)
16	Experimental	2(1)	4(0)	0	6(1)
⋮	Control	8(7)	1(0)	0	9(7)
17	Experimental	5(3)	4(0)	0	9(3)
	Control	8(7)	5(3)	0	13(10)
Totals					
Experimental		103(31)	81(18)	40(8)	224(57)
Control		115(54)	75(28)	32(6)	222(88)

[a]Vertical tie marks indicate pairs of instructors in the same college.

APPROACH TO THE PROBLEM

We chose to guide the analysis in terms of three questions.

1. Does the overall eight-week instruction program help? Does it help for all races? Does it help for both sexes?
2. Is coaching anything more than just anxiety reduction? How does its effect differ for different races? For different sexes?
3. Does anxiety reduction instruction exert effects beyond those which occur merely from having taken a previous test? For different races? For different sexes?

It is convenient in discussing these questions to let X represent the pretest score, Y the score on the anxiety reduction test, and Z the xcore on the post-test.

Then to address question (1) we examine differences of the type Z - X, where both scores come from the same experimental subject. For question (2) we look at differences of the type

Z - Y, where both scores come from the same experimental subject. Finally, to address the third question we turn to the control subjects, examining the difference Y - Z. This difference embodies the change following the anxiety reduction session in the controls.

We always use intrasubject differences, expecting the kind of advantage that matched scores provide. Complete cases can be used in any question they apply to, but in addition for various questions certain incomplete cases can also be used. The analysis employs every case for which the two scores in question were available.

We would expect the efficacy of treatment to depend upon the instructor (and other aspects of the instructional setting) as well as upon the student. Indeed, Table 2 shows very great heterogeneity of the classes with regard to completing the course. We should be prepared to find similar heterogeneity in outcomes, and so analyze the data in a way that takes account of that likely variability. This requires us at the least to examine the data class by class.

The two underlying approaches to the analysis then are: (1) employ intrasubject differences and (2) look at the data separately class by class.

ANALYSES AND RESULTS

Question (1): Does the Eight-Week Instruction Program Help?

The differences Z - X were calculated for all experimental subjects furnishing both scores. Table 3 presents the results, class by class. There are 15 classes with the required subjects. The variation in the means across the instructors is large compared to the variations within the instructors, so we disregard the within-instructor variance (as under-estimating the real uncertainty). We see that 12 of the 15 mean differences are positive, with class 14 (based on one subject) an extreme outlier and class 9 (based on three subjects) considerably far from the rest of the cases. The confidence intervals are shown in the table. The 96.4% sign-test-based confidence interval gives significant evidence of a positive median true effect. The 95% t confidence interval using all 15 instructors and weighting them equally (ignoring the variability in the number of subjects) does not give a significant result. The disagreement in the two results is due entirely to class 14 with one subject, and a very aberrant score. If this case is dropped, the t confidence interval using the remaining 14 cases again shows a significant result

TABLE 3. MEANS AND VARIANCES OF INTRASUBJECT DIFFERENCES Z - X
IN EXPERIMENTAL CLASSES

Instructor Number	Mean	Variance	Number of Subjects
1	0.012581	0.0018412	4
2 . . .			
3	-0.010213	0.011344	5
4 . . .	0.056568	0.021448	7
5	0.035356	0.0088164	3
6	0.063622	0.010494	8
7 . . .	0.053932	0.017486	12
8	0.044184	0.025712	13
9 . . .	0.13959	0.00015771	3
10	0.077434	0.0031882	6
11	0.034219	0.014671	7
12	0.064124	0.0060382	10
13	0.023538	0.011591	5
14 . . .	-0.17033		1
15			
16 . . .	0.0078933	0.0056242	2
17	-0.0028386	0.026072	3

96.4% Sign-test-based confidence interval for median overall
instruction effect: (0.0078933, 0.063622)

90

TABLE 3 (Continued)

Instructor Number	Mean	Variance	Number of Subjects

95% t Confidence intervals for overall instruction effect:

using all 15 instructors	(-0.0081273, 0.065415)
omitting case 14	(0.020770, 0.064942)
omitting cases 9 and 14	(0.018664, 0.052166)
omitting the five cases with three or fewer subjects	
	(0.022680, 0.061318)

at the 0.05 level; similarly, if the other outlier (case 9) is also dropped, or if all cases involving three or fewer subjects are omitted, the resulting t confidence intervals all show significance at the 0.05 level. Thus the ambiguity as to significance is not great. Only when the least well-determined and worst outlying data point is allowed to enter with full weight into a t confidence interval does the interval slightly overlap 0; otherwise significance at 0.05 is found. Hence we conclude that the overall instruction effect is significant, and that performance rises, on the average, between pretest and post-test for the experimental subjects--who received both anxiety reduction and coaching instruction.

The subsidiary question--does this effect differ between the sexes--is examined in Table 4. There we see that of the 17 experimental classes there were 13 in which both male and female subjects furnished scores Z - X; in six of these classes the differences favored the females, and in seven the males. We cannot see here any indication of a sex difference in response to coaching.

In a similar way we may examine the question of whether the gains Z - X of experimental subjects show a difference between races. Table 5 gives the averages for each race in each experimental class. We are not able to learn much from the data. For example, there were only five experimental classes involving a comparison of Chicanos with whites; of these five the difference favored the Chicanos in classes 6 and 8 and the whites in classes 3, 7, and 10. The black-white comparison could be made in only four classes, in showing greater average gains for the blacks in class 10 and smaller ones in classes 4, 13, and 16. Indeed, we

TABLE 4. DIFFERENCES BETWEEN AVERAGE Z - X SCORES FOR MALES AND FEMALES[a] IN EXPERIMENTAL CLASSES, TOGETHER WITH ESTIMATED VARIANCES OF THOSE DIFFERENCES

Instructor Number	Sex	Mean	Variance	Number of Subjects	Difference in Means	$V\left[\dfrac{\text{Difference in Means}}{}\right]$
1	M	0.012581	0.0018412	4		
	F					
2 · · · ·	M					
	F					
3	M	-0.097411		1	-0.10900	
	F	0.011587	0.011957	4		
4 · · · ·	M	-0.055680	0.013357	3	-0.19643	0.0074377
	F	0.14075	0.011942	4		
5	M	0.058839		1	0.035225	
	F	0.023614	0.016806	2		
6	M	0.12361		1	0.068562	
	F	0.055052	0.011558	7		
7 · · · ·	M	0.054343	0.013361	4	0.0006168	0.0060591
	F	0.053727	0.021751	8		
8	M	0.070853	0.010334	7	0.057784	0.0093347
	F	0.013070	0.047151	6		

TABLE 4 (Continued)

Instructor Number	Sex	Mean	Variance	Number of Subjects	Difference in Means	$\sqrt{\lvert \text{Difference in Means} \rvert}$
9	M	0.13820	0.00031250	1	-0.0020830	
	F	0.14028		2		
10	M	0.077260	0.0028801	4	-0.00052090	0.0043702
	F	0.077781	0.0073003	2		
11	M	0.12560	0.0094795	2	0.12794	0.0074980
	F	-0.0023343	0.013791	5		
12	M	0.12071	0.0099414	3	0.080835	0.0038076
	F	0.039874	0.0034565	7		
13	M	-0.097219	0.0093781	1	-0.15095	
	F	0.053727		4		
14	M	-0.17033		1		
	F					
15	M					
	F					
16	M	0.060922		1	0.10606	
	F	-0.045136		1		
17	M	-0.066065	0.028158	2	-0.18968	
	F	0.12361		1		

93

TABLE 4 (Continued)

Instructor Number	Sex	Mean	Variance	Number of Subjects	Difference in Means	V	Difference in Means

97.7% Sign-test-based confidence interval for median sex difference in overall instruction effect:
(-0.15095, 0.080835)

95% t Confidence interval for sex difference in overall instruction effect:
(-0.080861, 0.054454)

[a]Positive difference indicates larger male Z - X score.

94

have no basis here for positing racial differences in response to the experimental regimen.

We may observe that in addition to the four classes we have taken into account there were three classes which furnished only black data and eight which furnished only white data. To these 11 classes we can apply the Wilcoxon two-sample rank test; we find that the rank sum for the black classes is 15, slightly smaller than the expectation, which is 18. Since we have ranked from least to greatest this weakly corroborates finding only one of four classes in which the blacks had the greater gains. The two results could be formally combined, but obviously would still be insignificant at conventional levels, so we forego that exercise here.

Question (2): Is Coaching Anything More Than Anxiety Reduction?

We found in the analysis of Table 3 that the experimental regimen indeed increased scores, by about 2 to 6 percentage points. But the "experimental regimen" includes both an anxiety reduction phase and a later substantive coaching phase. Is there an effect of the substantive coaching beyond the level reached in the post-anxiety-instruction test? That is question (2). We examine it by looking class by class at all the experimental subjects where the difference $Z - Y$ can be computed. The results appear in Table 6.

There we see 11 classes in which the value of $Z - Y$ was observed for one or more students. In five of them the average difference was negative, and in six it was positive. Here there is no indication of reliable score improvement due to instruction. As might be expected, these fragmentary data fail to show race or sex differences when broken down as was done in Tables 4 and 5. These breakdowns are not shown here.

Question (3): Does Anxiety Reduction Instruction Exert Effects Beyond Those That Occur Merely From Having Taken a Previous Test?

We have found that the whole eight-week experimental regimen raises scores, and that this does not appear to be due to the substantive coaching. We now inquire directly into the effects of the anxiety reduction instruction. We could again use the experimental subjects, but prefer to turn to the control subjects for this question. The reason for this preference emerges from study of the schedule. Suppose in the experimental group we examined the scores $Y - X$ and found a positive difference (as we

TABLE 5. AVERAGE Z - X SCORES FOR EXPERIMENTAL SUBJECTS,
CLASSIFIED BY RACE. NUMBERS OF SUBJECTS SHOWN IN PARENTHESES

Instructor Number	White	Black	Chicano
1		.012581 (4)	
2 ⋮			
3	.011587 (4)		-.097411 (1)
4 ⋮	.072477 (2)	.050204 (5)	
5	.035356 (3)		
6	-.009815 (2)		.088102 (6)
7 ⋮	.096531 (5)		.023505 (7)
8	.034171 (8)		.060204 (5)
9 ⋮	.13959 (3)		
10	.069448 (4)	.14653 (1)	.040281 (1)
11	.034219 (7)		
12		.064124 (10)	
13	.099939 (3)	-.091065 (2)	

TABLE 5 (Continued)

Instructor Number	White	Black	Chicano
14		-.17033	
⋮		(1)	
15			
16	.060922	-.045136	
⋮	(1)	(1)	
⋮			
17	-.002839		
	(3)		

would expect, having found Z - X to be positive and Z - Y to be approximately zero). Could we conclude that the anxiety reduction session raised scores? Possibly. But such an increase could occur simply because taking the test for the second time leads to increase in scores--with or without anxiety reduction instruction. So we study Y - Z in the control group when the Z score is already a second test and we are looking at Y, a third test immediately preceded by anxiety reduction training. Will Y exceed Z on the average? (Recall that a third test immediately preceded by substantive coaching did not show a rise.)

The results appear in Table 7. There are 12 control classes with one or more students who furnish Y - Z scores. Two of these have negative average differences, and the other ten have positive ones. The median of the 12 class means lies between the scores for classes 11 and 12; the average of their values, our estimate of the median, is .059308. The confidence intervals enable us to conclude that there is a real increase in scores after exposure to the anxiety reduction instruction, and that this is not a "second test effect." We doubt it is a "third test effect" since no such effect appeared in the experimental subjects even in the presence of substantive coaching.

Breakdowns by race and sex failed to indicate any reliable differences in the main effect; the data are not presented here.

TABLE 6. MEANS AND VARIANCES OF Z - Y SCORES IN EXPERIMENTAL
SUBJECTS

Instructor Number	Mean	Variance	Number of Subjects
1	-0.10913	0.0071867	4
2 . . . 3			
4 . . . 5	0.020382	0.023286	7
5	-0.083176	0.0050465	4
6			
7 . . . 8	-0.047036	0.030128	12
9 . . .	0.088094	0.00097656	3
10	0.016219	0.033309	6
11	-0.011415	0.011321	7
12	-0.054445	0.065852	6
13	0.044063	0.012425	5
14 . . . 15			
16 . . .	0.190515		1
17	0.063901	0.019373	3

93.4% Sign-test-based confidence interval for median
increase in scores due to coaching: (-0.054445, 0.063901)

TABLE 6 (Continued)

Instructor Number	Mean	Variance	Number of Subjects

95% t Confidence intervals for increase in scores due to coaching:

using all 11 instructors	(-0.046918, 0.068367)
omitting case 16	(-0.053779, 0.039270)
omitting the three cases with three or fewer subjects	(-0.073265, 0.017130)

SUMMARY

We have found a rather surprisingly clear interpretation of these data:

1. There is a definite increase in scores, generally repro-
   ducible across classrooms, over the eight-week program.
2. This increase appears to occur early in the program,
   before the substantive instruction in quantitative reason-
   ing is introduced.
3. The increase in scores does follow upon the one session
   addressed to reduction of anxiety in the test-taking
   situation.
4. There was not enough information concerning possible sex
   and race-group differences to enable precise statement,
   but review of what evidence there was turned up no strong
   or even very suggestive indications of such differences.

DISCUSSION

The analysis of these data has proceeded largely by use of
confidence intervals based on the t test. Exception could be
taken to this on the grounds that the various classes furnish
unequal numbers of subjects and that it is incorrect to symmetrically treat two classes, say one with 11 cases and another with
three cases.

Despite the plausibility of this objection we do not find it
persuasive with these data, and would generally accord it not
much weight with data of the same sort. Our reasons are the following three:

TABLE 7. MEANS (AND VARIANCES) IN Y - Z SCORES IN CONTROL
CLASSES

Instructor Number	Mean	Variance	Number of Subjects
1	0.0029531	0.015338	5
2 . . . 3			
4 . . . 5	-0.0065627	0.0016835	5
5	-0.010014	0.014715	7
6			
7 . . . 8	0.036031	0.013633	15
9 . . . 10	0.10402	0.033342	8
10	0.093860	0.000019530	2
11	0.057243	0.014732	16
12	0.061373	0.024786	11
13	0.096633	0.010929	4
14 . . . 15	0.078235	0.017578	2
16 . . . 17	0.089094	0.024264	7
17	0.040070	0.013054	10

96.1% Sign-test-based confidence interval for median
increase in scores due to anxiety reduction:
(0.0029531, 0.093860)

100

TABLE 7 (Continued)

Instructor Number	Mean	Variance	Number of Subjects

95% t Confidence intervals for increase in scores due to anxiety reduction:

using all 12 instructors   (0.027396, 0.079759)

omitting the two cases with three or fewer subjects
                                    (0.016900, 0.077268)

1. Conclusions from the study would desirably be generalized to future--and other--classrooms, as well as to other students.   Then we must regard the classes in this experiment as a sample, which implies that our fundamental independent observations are <u>class</u> summaries. We want to assess any consistent tendencies in terms of their variability across classes.   It is unfortunate that the class averages indeed do have unequal variability, but we must live with that.

2. There are certain circumstances under which analysis of the means of unequal sized classes is "exact," or nearly so.

   a. If we were in the special case of normality with all classes of equal size (and within class variance σ_W^2 was the same for all classes), then analysis in terms of class means sacrifices no information at all if there is true class-to-class variation, and loses only in the <u>degrees of freedom</u> for the t test if there is no true class-to-class variation.

   b. As the between-class variability σ_C^2 grows large compared to the within-class variability σ_W^2, the difference in sample sizes matters less and less.   In a natural notation we have that

$$\sigma_{\bar{X}_i}^2 = \sigma_C^2 + \frac{1}{n_i} \sigma_W^2 \, ,$$

which can be written as

$$\sigma_{\overline{X}_i}^2 = \sigma_C^2 \left[1 + \frac{1}{n_i} \frac{\sigma_W^2}{\sigma_C^2} \right] ,$$

and it is clear that if σ_W^2/σ_C^2 is very small then it matters hardly at all whether $n_i = 1$ or 100.

c. If the class means are symmetrically distributed about a common mean but with unequal variances, then exact confidence intervals for that unknown common mean are arrived at by application of the sign test procedures used here-- and also by Wilcoxon-signed rank test confidence intervals. Both tend to agree with the t test.

3. It is not possible to give a "best" solution, except under fairly strong assumptions (such as normality of class means and of individual scores). And then such procedures are quite ponderous. We find that multiple analyses, omitting the least-well-determined values, or retaining them, lead to closely similar interpretations.

A point of probably greater weight in the confidence due this analysis relates to the effects of selection as students dropped out. To just what population of students (and classes) do our findings extend? That select group of students who would stick it out for three weeks (on some conclusions), or for eight weeks (on others)? Do such students differ in important ways from GRE takers generally? We can see in Table 2 that selection factors are strong and heterogeneous. For example, in the experimental groups complete cases were about 30% of the whites, but in the other two groups about 20%. In the control groups there is a similar difference. There is also a much higher complete-case ratio among the controls, but that is natural since a control case completed the last test at week 4 rather than week 8. In any case the data of this study relate largely to "survivors" of a high-attrition process. The degree to which conclusions will apply to other teaching situations is an important question to which answers must be sought outside the data of the study itself.

Added in Press. After this report was written the investigator discovered an error in his instructions for reading the tapes containing the data. Careful consideration indicated that effects induced by this error should be slight. Tables 3 and 7 were checked in detail with corrected data; effects were indeed

slight and no substantive conclusions were affected.   The conclusions of this analysis as a whole are believed to be unaffected.

A RETROSPECTIVE STUDY OF POSSIBLE CAUSALITY FACTORS

ROBERT WOLFE

Medical Problem. Investigating the factors associated with bron-
chopulmonary dysplasia (BPD) in new born infants.

Medical Investigators. William Northway, David Edwards, Thomas
Colby, and Wayne Dyer, Stanford University.

Statistical Procedures. t test; chi-square; multiple logistic
regression.

MEDICAL BACKGROUND

BPD is a respiratory problem that affects some newborn infants
who have had respiratory distress syndrome (RDS) and subsequent
oxygen therapy for it. BPD has also been observed in some
infants without RDS who have received high levels of oxygen or
intubation therapy. It is suspected that oxygen or intubation
therapy cause BPD. BPD is a deterioration of the lung tissue
evidenced, for example, by scarring. It can be diagnosed by a
physician by visual inspection of an infant's lung X ray or by
direct examination of lung tissue upong the death of an infant.
The severity of the deterioration is measured on a I through IV
scale, IV being the most severe. For the statistical analysis
presented here, infants reaching stage III or IV were classified
as having BPD, those reaching only stage I or II as not having
BPD.
All infants treated at Stanford Medical Center between 1962
and 1973 who had clinical and X-ray symptoms of RDS and who
received ventilatory assistance by intubation for more than 24

hours were included in the study, except for one whose records were unavilable. There were 299 such infants.

There were four types of data for each infant (see the Appendix for a listing). General background and health indices included the sex, year of birth, one-minute APGAR score, gestational age, birthweight, age at onset of symptoms, and severity of the initial X ray for RDS. The measures of the therapy given were the age at onset of ventilator therapy, the total number of hours of exposure to low (21-39%), medium (40-79%), and high (80-100%) oxygen, and the number of hours of endotracheal intubation. The medical outcomes for each infant included the highest stage of BPD reached as observed by X ray, the stage of BPD present at death as judged by direct examination of lung tissue for those who died, the number of hours of survival up to the end of the study, and whether or not the infant was alive at the end of the study. Finally, several pathological indices of the condition of the lungs of deceased infants were available.

STATISTICAL ANALYSIS

The investigation was open-ended but concentrated on four main questions:

1. Do oxygen or intubation therapy cause BPD?
2. How are the pathological indices related to BPD?
3. Which background or health variables indicates high risk for contracting BPD?
4. How serious is BPD to an infant's life?

One problem encountered in answering the questions was that of competing risks. Infants who may have gotten BPD died from other causes (e.g., RDS) before they contracted BPD. This makes comparison between the two groups, infants with and without BPD, difficult. Since BPD is a condition that requires time to develop, this problem was addressed by only including those infants who survived for three days or more in the analysis. This reduced the sample population size to 248, 170 without and 78 with BPD. Another possible approach would have been to exclude all infants from the analysis whose autopsies indicated death from a cause other than BPD, but such information was not generally available.

Question (1) was difficult to answer with the data available because oxygen and intubation therapy were given in amounts depending on the symptoms shown by the infants. In particular, low oxygen was often given as therapy for BPD and such therapy could hardly be considered as a possible cause for BPD! The medium- and high-level oxygen and intubation therapy were generally given before the severe stages of BPD were evident according to the medical researchers. Because of this entanglement between the symptoms and therapy, the implications of the analysis reported here should only be relied upon to show directions for further research. The results of the analysis of the relationship between BPD and oxygen therapy for these data are described next.

It was found that logarithmic transformations of hours of intubation and oxygen therapy gave more homogeneous variances and less skewed distributions among infants with and without BPD than did the untransformed therapy measures, so logs were used for the analysis. On the average, infants with BPD had received more hours of intubation (I), high oxygen (H), medium oxygen (M), and low oxygen (L) than had those without BPD ($p < .001$ for each t test) (see Table 1).

Since oxygen therapy was generally given in response to the early RDS symptoms shown, the association between BPD and hours of therapy was analyzed with statistical adjustment for initial RDS severity using a multiple logistic regression approach (Cox, 1970). The logit of the probability of observing symptoms of severe BPD prior to the end of surveillance was modeled as a linear combination of RDS severity (0-5 scale) and log hours of therapy. All three oxygen measures, L, M, and H, were significantly associated with BPD ($p < .01$) with this adjustment for RDS severity. Results indicated that RDS was associated with BPD ($p < .05$) except when considered in combination with high level oxygen ($p > .3$).

For each of the three levels of oxygen therapy a logistic analysis with oxygen, intubation, and RDS was performed to assess

TABLE 1. MEDIAN HOURS OF THERAPY FOR INFANTS SURVIVING BEYOND THREE DAYS WITH AND WITHOUT SYMPTOMS OF SEVERE BPD

Therapy	With BPD	Without BPD
Low O_2	438	64
Medium O_2	309	73
High O_2	90	28
Intubation	411	126

their combined association with BPD. In all three analyses, the
oxygen and intubation measures were positively and significantly
(p < .01) associated with BPD, while RDS was not significantly
related to BPD (p > .3).

Since medium- and high-level oxygen were suspected of being
causally related to BPD, a logistic model with those two measures
was fitted to the data. The model can be written as

$$\log[P/(1-P)] = \alpha + \beta_1 \log(\text{hours med } O_2 + 1)$$

$$+ \beta_2 \log(\text{hours high } O_2 + 1) \ , \tag{1}$$

where P is the probability of an infant having BPD symptoms by
the end of the study. The estimated parameters for this model
and their standard deviations are given in Table 2.

Question (2), the relationship between pathological indices,
which were all discrete, and BPD was studied by means of chi-
square tests. Many of the pathological indices were significantly related to the BPD stage, indicating that the condition
known as BPD may be one symptom of a more complex problem involving other indicators as well (see Table 3).

Question (3), the relationships between BPD and indices of
general health at birth, were studied by means of t tests for
continuous indices and chi-square tests for discrete indices.
When all 299 infants are included in the analysis, the general
pattern that emerges can be explained by competing risks. Good
health is associated negatively with having BPD among infants who
were living at the end of the study and is associated positively
with having BPD among infants who died. The competing risks
explanation is that among infants who died, the least healthy
ones died before they could contract BPD, while among those who
lived, those in better health were less likely to get BPD. When

TABLE 2.   PARAMETER ESTIMATES FOR MODEL (1) BASED ON DATA FOR
INFANTS SURVIVING AT LEAST 72 HOURS

Parameter	Estimate	Standard Error
α	-5.26	.68
β_1	.74	.11
β_2	.34	.08

TABLE 3. ASSOCIATIONS BETWEEN PATHOLOGICAL VARIABLES AND BPD.
THE NUMBERS IN THE TABLES ARE THE NUMBER OF INFANTS IN THE APPRO-
PRIATE CATEGORY WHO LIVED FOR THREE DAYS OR MORE AND THEN DIED;
THE PATHOLOGICAL MEASURES WERE UNAVAILABLE FOR SOME INFANTS, SO
THE TOTAL NUMBER OF INFANTS IN EACH TABLE VARIES FROM TABLE TO
TABLE. ALL FOUR PATHOLOGICAL VARIABLES WERE SIGNIFICANTLY
RELATED TO BPD AT THE .001 LEVEL

Bronchiola Mucosa			Pulmonary Interstitium, Alveolar Septa		
Health	Less Severe	Severe	Health	Less Severe	Severe
No BPD	46	15	No BPD	50	14
BPD	3	20	BPD	4	28
Chi-square = 24.2			Chi-square = 34.7		

Alveolar Infiltrate Outside Areas of Infection			Emphysema		
Health	Less Severe	Severe	Health	Less Severe	Severe
No BPD	31	24	No BPD	40	6
BPD	3	28	BPD	5	20
Chi-square = 16.2			Chi-square = 28.4		

only the 248 infants who lived for three days or more were
included, a pattern of association between poor health and higher
risk of BPD emerged (see Table 4).

Question (4), if infants surviving more than three days are
considered, the following table results:

Health	Dead	Alive
No BPD	67	103
BPD	40	38

with chi-square = 3.1.

TABLE 4. t TEST RESULTS FOR DIFFERENCES OF HEALTH INDICES BETWEEN INFANTS WITH AND WITHOUT
BPD FOR VARIOUS CATEGORIES OF INFANTS. THE 248 INFANTS WERE THOSE WHO LIVED FOR THREE DAYS
OR MORE. SIMILAR REVERSALS IN ASSOCIATION BETWEEN THE SURVIVORS AND DESCENDANTS WERE NOTED
FOR THE DISCRETE MEASURES APGAR SCORE AND RDS SEVERITY

| | All 299 Infants | | | | 248 Infants | |
| | Living | | Dead | | | |
Health Index	t Value	2-Tail Probability	t Value	2-Tail Probability	t Value	2-Tail Probability
Gestational age	-2.0	.05	2.4	.02	- .56	.58
Birthweight	-2.2	.03	1.9	.06	- .99	.33
Age of symptoms	-3.1	.00	.7	.47	-1.45	.15

109

The death rate is .39 among those infants without BPD and is .51
among those with BPD; these rates differ with significance
$p < .10$.

DISCUSSION

The primary motivation of this research was to investigate a
possible causal relationship between oxygen therapy and a severe
respiratory problem of newborn infants, BPD. A strong association was found between oxygen therapy and the occurrence of BPD.
Some qualifications concerning the data and the noted association
should be stated, however.

Since the data were from a nearly complete census of a defined
population of infants, many of the usual problems of retrospective sampling (Mantel and Haenszel, 1959) were not relevant to
this study. The question of generalization from this group of
infants was not examined but some of the characteristics of this
group of infants can be found from the data in the Appendix if
desired.

Analysis of the relationship between BPD and oxygen therapy
should allow for adjustment for other important factors. Adjustment for initial RDS severity is important since RDS is likely to
be associated with both the oxygen therapy and the final BPD
status for an infant. Adjustment for other characteristics may
have been appropriate as well, but was not carried out in detail.

The major problem encountered in drawing conclusions from this
study lies in the entanglement of the therapy given and the symptoms shown by an infant. Both the final BPD status and the therapy given may well be associated with the early condition of the
infant, a condition which may not be reflected in the initial RDS
severity data. Oxygen therapy may have been given as a response
to the deterioration of an infant's condition, a deterioration
which was likely to lead to BPD. Thus the association noted
between oxygen therapy and BPD could conceivably be just a spurious one. In response to this problem, the medical investigators
reviewed the patient records and found that high oxygen therapy
was generally discontinued before lung damage was observed, so
that the noted association could well be due to a causal
relationship.

Another problem encountered was that of censored data and competing risks. Some infants in the study had not been followed
until full recovery and some infants died from causes other than
BPD. For analysis, infants were classified as not having BPD if
they were followed and survived for three or more days without
exhibiting symptoms of severe BPD. Had they survived and been

followed, some of these infants may have gotten BPD later. A study utilizing the time to detection of BPD, death, or end of observation would allow for explicit use of the time of exposure to BPD for all infants in the study. Methods for such an analysis have been developed (Cox, 1972).

The association between oxygen therapy and BPD seen for this data is strong enough to warrant further study, with care being taken to avoid some of the limitations of this preliminary study.

REFERENCES

Cox, D. R. (1970). The Analysis of Binary Data. London: Methuen.

Cox, D. R. (1972). Regression models and life-tables, J. Roy. Stat. Soc., Ser. B, 34, 187-220.

Mantel, N. and W. Haenszel (1959). Statistical aspects of the analysis of data from retrospective studies of disease, J. Natl. Cancer Inst., 22, 719-748.

APPENDIX: DATA FOR BPD STUDY

Key to BPD Data

Columns	Data
1-3	Patient identification number, 1 to 300 (total of 299: there is no patient numbered 118). These are in chronologic order, by and large (by medical record number)
4	Sex: 1 if male, 0 if female
5-6	Year of birth (62 to 73)
7-8	One-minute APGAR score (0 to 10; clinically, 0 is the worst)
9-11	Estimated gestational age, weeks × 10 (clinical estimates such as "34-35" are thus scored as "345")
12-15	Birthweight, in grams
16-17	Age at onset of respiratory symptoms, hours × 10 (a value of 0 means symptoms began immediately at birth. A value of 5 means 30 minutes, etc.)
18-20	Age at onset of ventilatory assistance, hours
21-24	Duration of endotracheal intubation, hours
25-28	Duration of assisted ventilation, hours
29-41	Duration of exposure to elevated oxygen, hours 29-33: the lowest concentration, 22-39% 34-37: 40-79% 38-41: 80-100% (bearing in mind that humidified O_2 is never really 100%). These values are overall totals that the patients were exposed to during their hospitalizations, up to discharge or death
42	Survival as of May 1, 1975, 0 if dead, 1 if alive
43-47	Duration of survival as of May 1, 1975. In dead patients, sometimes the duration of survival may be less than the total oxygen or ventilator time; this is

112

Columns	Data
	an artifact of rounding off. For patient no. 1, the actual value is 110,376 hours
48	Subjective severity of RDS on initial chest X ray; 0 = no RDS seen, 5 = very severe disease
49-50	Highest stage of BPD reached radiographically × 10; patients who reached an intermediate stage are scored thus: I-II = 1.5(15), II-III = 2.5(or 25)
51-57	Pathological variables: 0 indicates missing data, low values indicate a less severe condition
51	Bronchiola mucosa
52	Pulmonary interstitium
53	Hyalin membranes
54	Alveolar infiltrate
55	Inflation pattern, emphysema
56	BPD stage based on lung tissue
57	Hematoidin

APPENDIX (Continued)

```
 1062  370294800 30   88   77  102   89    0199999010
 2062  360248003 20  209  204    7  233     10  2401103242330
 3164  310144600  7   39   38    0    4    410    453101121110
 4164  300235300  9   49   49    8   37    130    585102141120
 5164 8400320310 58   41   40   79   26  31194632110
 6164  320186000  1  653  644    0  625   1560  7962304403440
 7164 9  170105  49   40   40    0   12    790    912103131320
 8064  310141705 17  103  102    0    3   1190  1225201132122
 9164103502438 03 27   64   61  115  136  65192568210
10164 83602020 0 28  147  146  438  416  435192376340
11064  400253750 49   90   86   34  174  56191656210
12164  240 822 0  1   37   37    0    0    370    375101130010
13164  3602154 0 34   31   28   97  162  89191200110
14164103602260 0 32   35   32   56   47  132190792310
15164  3052013 0 31  222  222 12141515  3240  54483400403440
16164  240 935 0 79  236  225   30  103   1610  3231300403440
17165  3402098 0 40  137  133    8   11   1580  1775104343230
18065  4002384 0 46   42   39   52  155  144189808110
19065 33402055 0 36   28   25  142  119  24189520110
20065  2601091 0  0  922  918 1370296810830 54242400403440
21165  3402013 0 39  124  122  790  161  231188728240
22065  3102098 0  5  137  137 1142  157  131188008240
23165 83602367 3 25   26   26    0    2    490    511101131110
24165 32801361 0  3   46   46    0    0    500    505102131320
25165  3101318  33   84   84    5   68    490   1224103232220
26165 13201984 0  0   48   48    0    0    480    483202221120
27165  270 900   2   44   44    0    6    400    46520
28165  3402381 3 36   36   36    1    8    640    723102121320
29165  3402211 3 22  390  390    0  998   5510  1548340440 3440
30065  320150010 22   93   93  253   99  60185512115
31065  320170010 35  179  179 1395  799  244185512340
32165  2901400 0  0   50   50    0    0    500    505102121120
33165 83151871 3 26  163  163    1   68   1450  2143204203130
34165 93402041E0 55  411  307 13181724  3310  33603400403440
35065  4280 9E4 0 21   59   59    0    0    790    795101121110
36165103602208 45 19   40   40    3   31    370    702152121210
37065  3602466 0 48  158  153  195  108  206182560330
38065 7300116220 27  109  109    0   15   1210  1364103232120
39066 83201474 3 22  768  768    0  278   5130  7902304403440
40166  3101360 0  9  243  243    0    0   2530  2543253202032
```

114

```
410E6  3401332 0 28  62   62   13    3  740     901103212330
421GG 83601829 0 37 180  110  360  130 109180232410
431E6 23401247 0  8 609  609    0    0 6170    6175304403440
441GG 84002948 5 29 561  527  658  892 623179776340
451G6 93602268   20  75   75    0    6  890     951152131320
461GG 83401570 0  5  48   48    0    0  530     535102141320
470E6  320173610 10  26   26    4    1  300     353101111210
48166  4003969 0 15  48   48    0    7  560     633102032020
49166 53602693 3 29 268  268    5   19 2690    2961204433040
501E6 1270 920 0  4 100  100    0    5  990    1043102143020
511C6 3270 907 0  1  45   45    0    1  450     465102131320
521EG  3401616 0 16 283  283    3   29 2670    2993304202030
53066  3301750 0 10 186  186    1   10 1860    1965103202020
54066 13601240 0 38  45   45    0   10  730     833102131020
55166 83C03062 5 16 273  273    0    4 2860    2901303302331
56166 83552325 5 30 129  129    1   18 1400    1593102232222
57166 43502523 0 23 194  194    2   25 1900    2172204432440
58166 63401701 3 39  93   92  185  103 171173584230
591C7 93552069 5 59 493  493    2  102 4710    5752304403440
60067 5310 964 0 16 135  135    0   69  820    1513151200021
61167 7300 822 0 80 654  654   48  642 2200    9122300403440
62167 93151446 3 11  55   55    3    3  590     653202231220
63067  3652197 0 44 753  753    3  483 6590   11493404403440
64167 13302154 0  8 133  133    0    4 1370     141320
;6167 93C02268 5 46  44   44   46   90  97170152110
66067  3402410 0 27 149  149   90   83 123169792240
67167  3101460 0 19  30   30    0    8  410     485102110120
68067 73402268 0  7  26   26    1    1  310     332101131210
69067  300 990 0 53 185  185  626  225 60168832340
70067 7300 935 0  7  31   31    1   20  170     383101110110
71167 83202069 0 35 477  392  445  667 90168616440
72067  280 992 0    51003 882  842  611 66167872540
73067 93101860 0 16 441  372  892  517 57167800340
741C7 33301361 0  4 714  690    0  638  780    7174404403040
751E7 93702400 0 41 100   82   46   68 54167464310
76167 53401843 0  3  38   38    0    3  380     415101130010
77167 2250 897 0 79 309  250 1264  292 0166576130
78167  3201162 0  4 264  264    0  147 1210    2683204303430
79167 9   1686 2 18 199  199    0  200  120    2171100402440
80067 53502069 0 13 296  296    0   55 2550    310320
81168 7360262520 69 230  230    0   50 2370    2983304332230
82168 33101106 0 76 185  156 1955  190   00 24003300403440
85168 9350292010 16 835        1567  659 37161992340
34168 73302098 0 74  76   76   34   68 18161920310
85168 73602400 5 62 125  103   55  105 45161560310
36168  3802500 0 12  60   54    0    3  580     605151122010
87168  320235010 56 157  109  182   80  7161368320
88168 23301928 0 47 108   97   64  174 16159952310
89168 83302080 8 71 161   93   56  240  0157528320
90168 64002720 0  9 137   87   50  130 18157432310
```

APPENDIX (Continued)

```
 91068  3001502 0   2 263 263   45  198  220  2651304433042
 92168 43502240 0  13 440 326  580  145  58156400330
 93068  8290 652 0  11  46  46    2   23  320   563101141210
 94168 83101928 0   4  41  41    0   22  230   455102140220
 95168  1265 992 0  77 125  83 1278  431  21155752110
 96168 74003473 2  40 194 147   73  102  81155680230
 97169 52751134 0   8  92  92    0   17  830  1003102342322
 98069 13101260 0224 260  54  127  331   0155272310
 99069 63251360 0   3 320 318   20  263  410  3233204400040
100169 53602636 0   5 215 215    1  116 1020  220310
101069 92901140 5  24 105 105    0   24 1050  2133102142122
102069 7300154515 18 196 196    5   20 1880  2132253343332
103069 73001474 0  41 191 183    5  124 1030  2325253230331
104069  3202000 5  14 175 172   12  128  470  1861203243331
105069 8400331710 57  89  86   73  122  75151648310
106169 63102041 0  36 211 211    4   44 1980  2463204442141
107169 73001106 0  13 156 156   25  142   20  1681202340222
108169 53552013 5  21 269 269    0   17 2720  2905203232331
109169  4002438 0  58 471 320  230  496  50150760340
110169910365255210 14 148 101   12  146  63150208215
111169 52951219 0  15  47  47    1   23  400   633102131221
112169 63401645 0  21 158 103  318  100  18149224310
113169  330201310 12  31  31    0    0  390   42410
114169  3352050 0  38 290 239  102  230  91148240530
115069 73101361 0  27 627 577  320  640  1147400130
116169 73251090 5  38 491 472  711  195  4147232330
117169 13652622 0  510791079    0  998  860 10843304403440
119170 65503058 5   7 194 193    1   16 1850  2025203443042
120170 5310124610 25 173 166   76  102  210  1983103333231
121170 53401540 0  19 193 193    0   73 1420  2153204302331
122070 9290155425 16 353 350  117  176  800  3733103403032
123170 93151500 3  10  30  30    0    8  320   393102233222
124170 83201400 5  66  76  72   33  102  5144160310
125170 63301786 0  39  96  91  122  115  23144112310
126170 83702268 0  35 796 792    4  647 1830  8832303303030
127070  3280 960 0108 722 688 1415  723  6143320340
128070 72901304 0   8 258 258    0    6 2610  267320
129070 53302112 0   9  90  90    2   20  770   983153231 30
130170 53602792 0111 402 402   73  309 1400  5203304403440
131170 9345252330 34  79  76  102   99  40141232110
132170 7380317520 83  89  89   74  122  41141160010
133170 83001320 0  12 140 139    2    7 1420  1525103343330
134070  2701000 0  11 192 192    0   58 1390  1981104343332
135170 33452154 0  10 209 209    0    5 2170  2213204433041
136170 33201899 0  14 221 219    0  144  900  2335203432130
137070 73602340 0  21  51  50  120  106  11140272110
138170 53101701 0  17 767 677 1104  711 2690 32883404403440
139170 53401786 0  26 274 266 1110  731  71139432140
140170 93502551 5  46  65  59   70   75  20138952310
141170  1260 992 0  35  74  74    0    0 1070  1071102231022
142170 12701300 0   1  56  56    0    0  570   571101131112
143170 33302220 0   5  29  29    0    3  300   335102110010
144170 83201843 0   3  54  54    0    5  520   57110
```

116

```
145170 93201843 5 51   79   79     2   12 1170  1303102132022
146170  3602750 5 20  354  334   703  220  66138112230
147171 13301786 0 13  243  243     0   46 2090  2551253433441
148171 43101250 0  6   41   41     0    2  450   47310
149171 83602572 5 57   57   54    84   82  18137032310
150171 73501360 5 25  138  110  1155  222  29136864340
151171 73552551 0 33   65   61     9  110  17136816110
152171 8365255110 64  63   52   101   58  20136744110
153171 6290122020 48  226 184   275  128   4136408310
154071 95401644 0 63   54   54     2    1 1140  1173102130010
155171 33302040 0 34   65   61    78   83  51134968310
156171 6310140010 31  100   80   739  114  30134848530
157171 83101276 0 14   52   52    86   24  10134704110
158171  325171510 10  , 6   63     0    5  680   73310
159171 13452325 3 12  138  138   324   84 199153912240
160171 25201984 0  9   42   42     5    1  460   525102221220
161171 6310136010  8  132  129     0   59 1130  1725104233041
162071 13201600 0  3   58   58     0    4  580   625103242220
163171 33001729 0 40  242  185    23  171  740  2832202203322
164171 42901332 0 38  578  313  1120  120  200  23041100403040
165171  2901220 5  22 103  103     1    3 1220  1265203243232
166071 73201531 5  8   48   48     0    0  560   575102131221
167071 0510155910 15 182  179   430  197  16153120310
168171 4350236710 11 637  580   134  533  710   741340
169171 63201928 0 68  192 18911106  652  86029688340
170071  4280 851 5 10   24   24     0    4  300   344102130120
171171 83201786 0 47  280  132   290  120  67132760310
172171 83202211 2  3  190  190     0   40 1500  1903204333032
173071 53301928 0 20  123  118   902   91  26132232340
174171 53001360 0 1223452012  1901  367 211131968230
175171 13501984 0 16  244  222   475  132  63151656220
176171 73501814 0 44  158  133   113  175   2131512110
177071  3602530 0 15  241  224   170  246 122131392430
178071 8370269310 1112691269   633  451 1630  1247540
179071 1300 851 0  .1  47   47     0   20  280   483101120210
180171 9360201315 45 128  104    65  168  12130720110
181171 13001077 0  1  143  142     0    0 1430  1433102343332
182171 73402013 0 13  196  164   294   69  43129760315
183171 12951200 0  3  234  234     0   59 1800  2383204131021
184071 8370272220 56 107   72    86   92  21129688310
185171 8360284510 29 194  140    81  112  72129640310
186171 8360275010 35 138  110   246  132  42129640310
187071 6360158710147 134 126   203  218   40  4243100403440
188172 63001247 5 12  205  205     0   80 1370  2164203432442
189172 93352410 2 49  123  108    17  170  24128272310
190172 83252260 0 18  264  261    84   9f 119128248340
191172 63201587 0 25  771  748    97  619  810  7973304403440
192072 83602183 0 28  552  497   720  434  6f127816340
193172 33201400 0  8   28   28     0   11  220   373102131220
194072 0280 992 0 711006 961  1336  193   00 1968130
195072 2801075 0 12   28   20     8   24   60   403103131220
196072 5270 950 0  5  236  231   223   58  7126856310
197172 12801050 0  9  218  216     0   60 1340  2143103342332
```

```
198072 34002395 0   7   27   26     0   5   280     343101021010
199172 73201460 0 141016  983   800 868 1560 19443300403040
200172 33401644 0   9  237  198   193  63  14126520310
201072 23301540 0  16   99   81    67 103  14126232210
202172 73401843 0   5  171  171     3 119   560  1695103233131
203172 43201729 0  13  140  111   201 122  16125512315
204172 85101220 0  29  154  154     0  98   840  1823153332132
205172 83101247 3  20   69   69     0  23   670    894102131211
206172  3602977 0  32  106  106   134  47  59124936310
207172 6320189930 49   92   92     0  83   540   137520
208172 53201389 0   8  185  184     3 156   340  1932204433042
209172  3401673 5  47   88   73   143  37   5124552310
210172 53002090 3  68  679  659  14803640 1020 52083400403440
211172 93201588 0  65  523  523   149 841 1860 11763403403330
212072  380255130 29 121  107    64 119  25124048110
213072  3401106 0  310761042  66621747  77011803404403340
214172 03101570 0  31  176  126   886  89   390 3232340
215172 84003161 0  27  211  193   145  53 111123016 20
216172 93001502 0   6   63   63     0  26   430    695102131210
217172 63602296 0  34   8 ( 66    57  71  12122536210
218072 43401446 3  26   92   91   125  87   5122536410
219172 53301300 0 3611541150   236 823 1000 11763404400040
220172 93802835 20 28   85   79    28  17  35122416310
221172 22901415 0   1   92   92     0  48   480    961103230322
222072 23401670 0  29  173  154   437  92  17121696310
223172 33601928 0  11   48   48     0   4   550    582102110121
224172 23201770 0   3  178  161    94 170  16121552440
225172 13201106 0   4   49   49     7  36   100   533103232321
226172 9370343753 38  42   42    39  53   7121360210
227172 1255 900 0   8   42   42     0  32   170   505103232322
228172 94003147 5  46  106  103    98  55  26120784320
229072 43201559 0  20  228  215   231  59  17120544310
230173  3602438 0  10  165  162   207  52  47120064210
231175  340206015 14 252  244  12731359  350 37444404403440
232073  340161015 33 162  153   248 160  22120016330
233173 53602637 5  38   72   71   106  35  13119992310
234173 83652240 5  26  118  118   121 120  26119848210
235173 73752680 3  22  127  127   163 106  12119824110
236073  4280 930 0  23   78   78   428  53  19119752310
237073 23051077 0   6  111  111    11  93   120  1173103130332
238073 83703900 0  23   99   99    54  97  15119152310
239173  350214030 32 126  124   179 106  31118864110
240075 32901100 0  28  113  109   108  23   8118408310
241173 73552714 2  28  203  202   134 134  11118312210
242173 53802792 0  45  151  151   104 154  25118192110
243173 83001360 3  30   51   51   100  47   0118144110
244173 7300124710 16 748  747  1513 486  15113144340
245173 83051180 0   8   28   28     0  22   140    362101131310
246175 52051175 0   8   63   63   727  77   6117760310
247073 73402126 0  48   57   57   183  43  24117496110
248173 7285109130  8  49   48     0   9   480    57310
249173 73501956 5  20   69   69    43  50  16117352310
250073 9300124710 54 420  412   143 436  19117280110
```

```
251073 13001205 3 49  63  63    3  75  310  1033102232222
252173 8330158720 19  35  35   88  39       3116992210
253173  3251701 0  7 137 119  153  72      16116992310
254173 73604026 3  4 221 219  665 200      24116704210
255173 9350224010 17 226 226 1801 537   18016680440440434440
256073 63401899 0 30 140  91   45 105       4116440110
257073 7360293420 38 123 123  103  63      17116152310
258173 1320 800 0  5 143 143    0  62  870  1483103142121
259073 8360222640 29  63  63   57  60      11115912210
260173 7330192810  8  98  98  175  87      11115864210
261173 73001587 0 1214361436  622 973      10115744240
262173 63551545 5 12  42  42    3  42  100   553102221120
263173 53401573 0  9 141 141   51  81  200    151320
264173 43201360 0 71 720 672 1329 405      13115744330
265073 73001106 0 21  31  31    2  36  120    512101110110
266173 53001106 0  713981398  512 886  160  13923404403440
267073 43351920 0  6 102 102   86  79       8115000110
268173 03201200 0  0  26  26    0  13  120    263102130010
269073 43101247 0  3 104 101  205  59      26114808410
270073  3401672 0 20  99  94  108  35      25114664320
271173  270 794 0  6  36  35    0   2  400   415102141220
272173 8375299010 25  86  86   72  48      11114616210
273173 83703062 5 11  91  91   98  70       1114592210
274173 43602100 0  6 160 160  160  4 \     14114232510
275173 33001420 0  1 240 240  827  79       5114112310
276173 8400290620 30  92  90   76  60       9114112210
277173 93051590 3  4 145 145  173  56       4114040310
278173 93102041 3 41  64  64   78  37       5113944310
279173 7340182015  5 861 821 1136 112       7113872315
280073 6290121910 35 237 230  635  64      10113704310
281173  3502596 5 29  98  98   93  32      25113656120
282173 83101318 5 46 840 839  817  44       0113512130
283173 23602700 0 14 132 132  137  41      41113440310
284073 32901450 0  4 147 147    0 113  370  1495203230331
285073 52801350 0 11  28  28    4  19  160   392101130310
286073 23001160 0  3  37  37    0  27  120   383102111320
287173 72851219 0 18 427 427  745  81  19113176415
288173 33402500 0 40  33  33   64  50       7113104110
289073 6265 970 0 21 197 197    2 151  660  2192204343231
290073 63602340 0 23 140 140   22 132  18112840210
291073 62801035 0 55  49  49   36  67   20  1041102131320
292173 22801220 0 13 156 156   85 100  10112504310
293173 53101550 0  3 456 421  395  16  17112288310
294173 33301588 0 42 222 196   75  94   9112240210
295173 43001230 0  4 507 507  338 167   50  5104304403041
296173 33001020 0  3 546 546  343 192  130  5475300403241
297073 6240 680 0 40 795 795  947  50  19111952110
298073 2260 750 0  4  25  25    0  21   80   283101130210
299073 2280 793 0  6 126 125    0  98  320  1305151130110
300173 2280 990 0  8  38  38    0  33  120   453103232322
```

PART TWO

SETTING DOSE LEVELS FOR THE TREATMENT OF TESTICULAR CANCER

BYRON WM. BROWN, JR. AND
MARIE S. J. HU

<u>Medical Problem</u>. Setting dose levels in the treatment of testic-
ular cancer so as to avoid serious toxicity.

<u>Medical Investigators</u>. John Daniels and John Krikorian, Stanford
University.

<u>Statistical Procedures</u>. Estimating dose-response curves from
binary data by nonparametric and multiple logistic methods; testing for goodness of fit.

BACKGROUND AND PROBLEM

Recently, encouraging results in the treatment of testicular
cancer have been obtained by the use of combinations of drugs
following surgery. The success of this chemotherapy depends on
the use of high dose levels of the drugs, dose levels that are
barely tolerable by some patients because of various forms of
drug toxicity. It would be useful to the chemotherapist to be
able to anticipate the thresholds for toxicity in the patient so
that the threshold could be approached without being exceeded,
thus giving each patient the maximum dose tolerable without
severe toxicity. The toxicity of concern here is gastrointesti-
nal (GI) and can be severe enough to require hospitalization.
At Stanford University Medical Center, Dr. John Daniels, Dr.
John Krikorian, and their colleagues treated 14 patients with
testicular cancer, each patient receiving one to six courses of a
combination of three drugs (cis-platinum, velban and bleomycin).
The doses of the first two drugs were varied from patient to

123

patient, and from course to course for a given patient, depending on the toxicity observed to that point in that patient and in previous patients.

Daniels and Krikorian tabulated the data on the 14 patients, for each of 55 courses received, indicating the doses of the drugs, certain variables that they thought might be predictive of whether the patient would experience serious toxic effects, and, of course, whether indeed, the patient did experience GI toxicity serious enough to require hospitalization following that course of therapy.

The data are shown in Table 1. The first two columns identify the patient and the series of courses given to each patient. The days between courses are given in the next column. This interval depended on scheduling and, in certain instances, on whether the patient seemed fit for the next course. The Karnofsky score is a widely used rating of the general physical well-being of the patient. The score ranges from 0 to 100% and the points on the scale are defined in Table 2. The patient's weight and body surface are given in the next two columns of Table 1; the latter was estimated only once since it remains quite stable. In the next three columns, the doses of platinum and velban, and use of bleomycin, are given for each course. Finally, in the last column the occurrence of serious, hospitalizing GI toxicity is recorded for each course.

The investigators had several questions with regard to these data. They believed that the velban was the cause of the toxicity, but when they calculated doses per square meter of body surface and attempted to correlate this with toxicity, they found only a weak relationship. Then they calculated the dose per kilogram of body weight and found a stronger relationship. Since velban doses are generally set by the body surface (and then modified in the light of toxicity), the stronger relationship to body weight was of interest, and they wanted a statistical judgment on its validity. The second, more general question was whether other variables, such as days between courses, could be used to predict toxicity in some sort of multivariate approach to the problem.

STATISTICAL ANALYSIS

Relationship of Velban to Toxicity

To demonstrate the relationship between toxicity and velban dose, computed relative to body surface and also relative to body weight, the investigators calculated the dose by each method and

TABLE 1. DOSE AND TOXICITY DATA FOR 14 PATIENTS RECEIVING 55 COURSES OF CHEMOTHERAPY FOR TESTICULAR CANCER

| | | | | | | Chemotherapy | | | |
Patient	Course	Days Since Last Course	Karnofsky Score	Weight (kg)	Surface (m^2)	Platinum (mg)	Velban (mg/day)	Bleomycin	Serious Toxicity
A	1	–	80	67.2	1.80	54	13.5	N	Y
	2	30	80	65.1		108	13.5	N	Y
B	1	–	70	65.3	1.90	190	13.0	Y	Y
C	1	–	70	60.3	1.75	180	12.0	Y	Y
	2	34	90	61.3		180	6.8	Y	N
	3	33	90	60.5		175	13.2	Y	Y
	4	41	90	65.3		175	8.0	Y	N
	5	28	90	64.7		175	8.0	Y	N
	6	28	90	65.7		175	8.0	Y	N
D	1	–	90	74.4	1.93	200	15.0	Y	Y
	2	38	80	71.1		195	14.5	Y	Y
	3	35	80	70.1		200	15.0	Y	Y
	4	38	80	69.4		200	10.0	Y	Y

125

TABLE 1 (Continued)

Patient	Course	Days Since Last Course	Karnofsky Score	Weight (kg)	Surface (m^2)	Chemotherapy			Serious Toxicity
						Platinum (mg)	Velban (mg/day)	Bleomycin	
E	1	–	80	59.6	1.65	185	13.0	Y	Y
	2	33	80	52.3		160	10.0	Y	N
	3	31	90	55.6		160	10.0	Y	N
	4	27	90	54.7		160	10.0	Y	N
	5	25	90	56.0		160	10.0	Y	Y
	6	33	60	51.9		160	9.0	Y	N
F	1	–	90	81.6	2.00	200	15.0	N	N
	2	29	90	83.0		175	15.0	N	N
	3	31	90	82.0		200	15.0	N	N
	4	28	90	84.0		180	12.0	N	N
	5	115	90	84.2		200	7.5	N	N
G	1	–	80	69.9	1.85	185	9.2	N	N
	2	28	80	67.8		180	10.0	N	N
	3	28	80	67.7		180	12.0	N	Y
	4	31	90	66.4		180	12.5	N	N

TABLE 1 (Continued)

Patient	Course	Days Since Last Course	Karnofsky Score	Weight (kg)	Surface (m^2)	Chemotherapy Platinum (mg)	Velban (mg/day)	Bleomycin	Serious Toxicity
	5	40	90	68.5		72	12.5	N	N
	6	48	90	70.0		144	10.0	N	N
H	1	–	100	74.0	1.96	190	14.0	Y	Y
	2	48	100	73.1		190	9.0	Y	N
	3	30	100	69.7		190	9.0	Y	N
	4	40	100	72.0		190	9.0	Y	N
	5	35	100	71.5		192	9.0	N	N
I	1	–	80	68.3	2.00	240	15.0	Y	Y
	2	38	80	67.3		190	14.0	Y	Y
J	1	–	100	101.2	2.26	225	16.0	Y	N
	2	28	100	100.6		225	16.0	Y	N
	3	28	100	98.4		225	15.0	Y	N
	4	27	100	97.5		225	15.0	Y	N
	5	70	100	99.5		225	17.0	N	N

127

TABLE 1 (Continued)

Patient	Course	Days Since Last Course	Karnofsky Score	Weight (kg)	Surface (m^2)	Chemotherapy Platinum (mg)	Velban (mg/day)	Bleomycin	Serious Toxicity
L	1	–	100	70.4	1.80	180	13.0	Y	N
	2	23	100	70.0		180	13.0	Y	Y
	3	77	100	70.3		150	11.0	Y	N
	4	43	100	70.3		120	11.0	Y	N
	5	38	100	70.3		120	11.0	Y	N
M	1	–	100	71.8	1.82	180	13.5	Y	Y
	2	23	100	69.5		180	13.5	Y	N
	3	27	100	66.5		180	13.5	Y	N
N	1	–	70	55.4	1.65	165	12.0	Y	Y
O	1	–	90	83.0	2.00	195	15.0	Y	N
	2	27	90	78.5		150	10.0	Y	N
	3	36	90	77.0		195	12.0	Y	N
	4	21	80	77.0		200	12.0	Y	Y

128

TABLE 2.   DEFINITION OF POINTS ON THE KARNOFSKY PERFORMANCE SCALE

100%	Normal, no evidence of disease
90%	Able to carry on normal activity, minor symptoms or signs of disease
80%	Normal activities with effort, some symptoms or signs of disease
70%	Cares for self, unable to carry on normal activity or do active work
60%	Requires occasional assistance but is able to care for most of needs
50%	Requires considerable assistance and frequent medical care
40%	Disabled; requires special medical care and assistance
30%	Severely disabled, hospitalization is indicated although death not imminent
20%	Very sick, hospitalization necessary; active supportive Rx needed
10%	Moribund, fatal
0	Dead

graphed the results.   The calculations are shown in Table 3. Note that the doses, per square meter and per kilogram of body weight, shown on the left-hand side of the table are for the 36 courses of treatment that did not have associated serious toxicity; on the right-hand side of the table are shown the doses for the 19 courses that were followed by serious toxicity.

The means, standard deviations, and standard errors of the means are shown at the bottom of Table 3.   It can be seen that

TABLE 3. VELBAN DOSE, CALCULATED PER SQUARE METER OF BODY
SURFACE AND PER KILOGRAM OF BODY WEIGHT, AND RELATED TO TOXICITY
FOR 55 COURSES OF CHEMOTHERAPY

Velban Doses (mg/day) for Courses Without Toxicity			Velban Doses (mg/day) for Courses With Toxicity		
Patient	per m^2	per kg	Patient	per m^2	per kg
C	3.9	0.11	A	7.5	0.20
C	4.6	0.12	A	7.5	0.21
C	4.6	0.12	B	6.8	0.20
C	4.6	0.12	C	6.9	0.20
E	6.1	0.19	C	7.5	0.22
E	6.1	0.18	D	7.8	0.20
E	6.1	0.18	D	7.5	0.20
E	5.5	0.17	D	7.8	0.21
F	7.5	0.18	D	5.2	0.14
F	7.5	0.18	E	7.9	0.22
F	7.5	0.18	E	6.1	0.18
F	6.0	0.14	G	6.5	0.18
F	3.8	0.09	H	7.1	0.19
G	5.0	0.13	I	7.5	0.22
G	5.4	0.15	I	7.0	0.21
G	6.8	0.19	L	7.2	0.19
G	6.8	0.18	M	7.4	0.19
G	5.4	0.14	N	7.3	0.22
H	4.6	0.12	O	6.0	0.16
H	4.6	0.13			
H	4.6	0.12			
H	4.6	0.13			
J	7.1	0.16			
J	7.1	0.16			

130

TABLE 3 (Continued)

Patient	Velban Doses (mg/day) for Courses Without Toxicity per m^2	per kg	Patient	Velban Doses (mg/day) for Courses With Toxicity per m^2	per kg
J	6.6	0.15			
J	6.6	0.15			
J	7.5	0.17			
L	7.2	0.18			
L	6.1	0.16			
L	6.1	0.16			
L	6.1	0.16			
M	7.4	0.19			
M	7.4	0.20			
O	7.5	0.18			
O	5.0	0.13			
O	6.0	0.16			
Mean	5.98	0.154		7.08	0.197
Standard deviation					
	1.16	0.028		0.70	0.021
Standard error					
	0.19	0.005		0.16	0.005
N	36	36		19	19

the courses that did lead to toxicity do indeed show higher
levels of velban than those courses that were not followed by
toxicity. This is true whether the dose is calculated per square
meter or per kilogram. However, when the distance between sample
means is calculated in terms of a pooled standard deviation
(Mahalanobis distance, in one dimension), the results are as
follows:

Dosage Method	Mean Dose Nontoxic	Mean Dose Toxic	Pooled Standard Deviation	$\left(\dfrac{\text{Difference in Mean}}{\text{Standard Deviation}}\right)$
Per square meter	5.98	7.08	1.03	1.07
Per kilogram	0.154	0.197	0.026	1.65

Thus the separation is more than 1½ standard deviations, using body weight as the basis for dose calculation, compared with a separation of only slightly more than one standard deviation, when dose is calculated per square meter. The investigators graphed histograms for the four groups, but a clearer display is achieved by plotting the cumulative frequency distributions, as in Figure 1. The greater vertical distances between the two cdf's, when the velban dose is calculated per kilogram of body weight instead of per body surface, seem clear in the figure.

From the point of view of the clinician, a more useful way of summarizing and graphing the data is to construct a dose-response curve for each way of calculating dose. The probability of a treatment resulting in toxicity is regarded as an increasing function of the dose level. This function can be estimated, as shown in Tables 4(a) and 4(b). As shown by Ayer et al. (1955), the maximum likelihood estimate of the dose-response function, under the constraint of monotonicity, is obtained by successive pooling of data from adjacent dose levels until a nondecreasing sequence of ratios is obtained. The results in Tables 4(a) and 4(b) are shown in Figure 2.

If it were true that the velban dose calculated on the basis of body weight is more predictive of toxicity than when calculated on the basis of body surface, then the dose-response curve for body weight should rise more sharply relative to a given percentage increase in dosage. The range (distance from zero percentile to 100th percentile) divided by the median does not show this, both ratios being about 40%, but the lower figure, except for the long left tail, does suggest a steeper rise.

On the basis of Figures 1 and 2 it would seem that the higher the dose level of velban, the greater the risk of toxicity. Furthermore, there seems to be some slight advantage in prediction of toxicity when the dose is calculated by body weight rather than by body surface, the usual method.

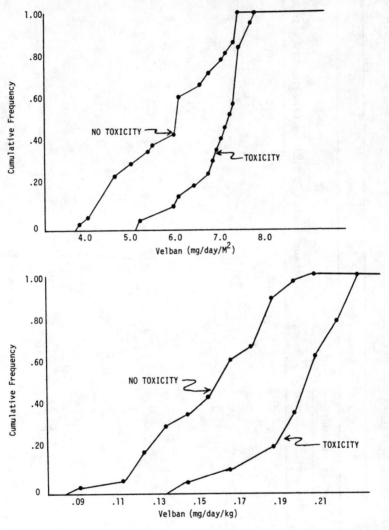

Figure 1. Discrimination between toxic and nontoxic courses of treatment based on velban dose, calculated per square meter and per kilogram.

133

TABLE 4. MAXIMUM LIKELIHOOD ESTIMATION OF THE PROBABILITY OF TOXICITY AS A MONOTONIC FUNCTION OF DOSE LEVEL OF VELBAN PER DAY

(a) Per Square Meter of Body Surface

Velban Dose (mg/day m^2)	Number of Courses		Observed Fraction Toxic	Maximum Likelihood Estimate
	Nontoxic	Toxic		
3.8 - 3.9	2	0	0/2	0/2 = .00
4.0 - 4.1	-	-	-	
4.2 - 4.3	-	-	-	
4.4 - 4.5	-	-	-	
4.6 - 4.7	7	0	0/7	0/7 = .00
4.8 - 4.9	-	-	-	
5.0 - 5.1	2	0	0/2	0/2 = .00
5.2 - 5.3	0	1	1/1	3/14 = .21
5.4 - 5.5	3	0	0/3	
5.6 - 5.7	-	-	-	
5.8 - 5.9	-	-	-	
6.0 - 6.1	8	2	2/10	
6.2 - 6.3	-	-	-	
6.4 - 6.5	0	1	1/1	1/3 = .33
6.6 - 6.7	2	0	0/2	

TABLE 4 (Continued)

Velban Dose (mg/day m^2)	Number of Courses		Observed Fraction Toxic	Maximum Likelihood Estimate
	Nontoxic	Toxic		
6.8 - 6.9	2	2	2/4	2/4 = .50
7.0 - 7.1	2	2	2/4	2/4 = .50
7.2 - 7.3	1	2	2/3 ⎫	8/16 = .50
7.4 - 7.5	7	6	6/13 ⎭	
7.6 - 7.7	–	–	–	
7.8 - 7.9	0	3	3/3	3/3 = 1.00

(b) Per Kilogram of Body Weight

Velban Dose (mg/day kg)	Number of Courses		Observed Fraction Toxic	Maximum Likelihood Estimate
	Nontoxic	Toxic		
0.09	1	0	0/1	0/1 = .00
0.10	–	–	–	
0.11	1	0	0/1	0/1 = .00
0.12	5	0	0/5	0/5 = .00
0.13	4	0	0/4	0/4 = .00

TABLE 4 (Continued)

Velban Dose (mg/day kg)	Number of Courses		Observed Fraction Toxic	Maximum Likelihood Estimate
	Nontoxic	Toxic		
0.14	2	1	1/3	
0.15	3	0	0/3	2/14 = .14
0.16	6	1	1/7	
0.17	2	0	0/2	
0.18	8	2	2/10	2/10 = .20
0.19	3	3	3/6	3/6 = .50
0.20	1	5	5/6	5/6 = .83
0.21	0	3	3/3	3/3 = 1.00
0.22	0	4	4/4	4/4 = 1.00

136

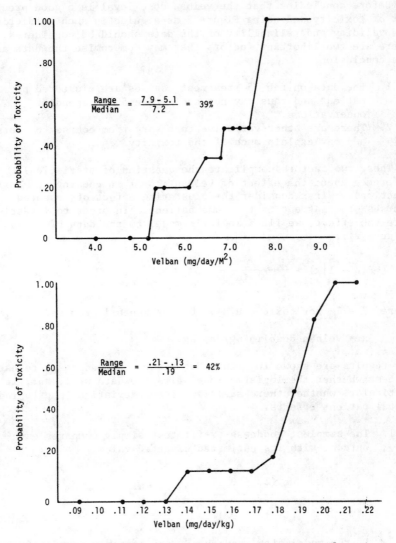

Figure 2. Dose response curves showing probability of toxic course versus velban dose calculated per body surface and per body weight.

Testing For Patient Heterogeneity

Before concluding that the velban dose level is a good predictor of toxicity and using Figure 2 as a guide in such prediction, the validity and reliability of the data should be considered. There are two important factors that may compromise the data and the conclusion:

1. The data on the 55 treatment courses are clustered by patient and thus may not constitute 55 independent observations.
2. There are other variables that vary from course to course and may explain much of the toxicity.

These two factors complicate the question of statistical inference about the effect of velban itself in causing the toxic reactions. First consider the clustering effect of repeated courses of treatment in the same patient. In order to investigate the effect, we fit a logistic model to the data of Table 3 (Figure 2):

$$P(Z = 1|x) = \frac{1}{1 + e^{-(\alpha+\beta x)}}$$

where $Z = 1$ for toxic course, 0 for nontoxic course,

x = velban dose in mg/day kg.

The results are shown in Figure 3. Several procedures were used to test whether the logistic model fits the data well, and, in particular, whether there are significant deviations due to individual patient effects.

1. The simplest goodness-of-fit test simply compares each outcome with its estimated expected value:

$$T = \sum_{i=1}^{55} \frac{(Z_i - \hat{P}_i)^2}{\hat{P}_i(1 - \hat{P}_i)} . \tag{1}$$

If the model fits, and the P_i are suitably bounded from zero and one, the asymptotic distribution of T is chi squared with 53 degrees of freedom. However, this test will have poor power and is not a consistent test against

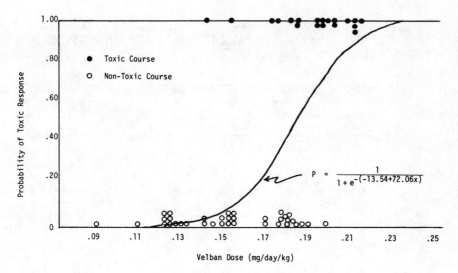

Figure 3.   Dose response fitted logistic curve showing the probability of toxic course versus velban (mg/day kg) and the observed
responses.

the alternative hypothesis of individual patient effects.
Calculation yields a value of 57.7 for T, and thus this
test might lead to misplaced faith in the logistic model
with common α and β for all patients.

2.   A test that is consistent against the alternative of individual patient effects would compare the results for each
patient against the expected number of toxic courses for
the patient, based on the fitted model shown in Figure 3.
The results are shown in Table 5.   The test statistic T',
defined at the bottom of the table, will have a distribution under the null hypothesis that is approximately chi
squared with 12 degrees of freedom; T' = 12.4787 is not
significant.   Thus by this test the data show no sugges-
tion of individual patient effects.

3.   A third approach to testing the hypothesis of a common
logistic function against the alternative of real patient
differences would be to fit the model

TABLE 5. CHI-SQUARED TEST COMPARING THE RESULTS FOR EACH PATIENT WITH THE EXPECTATION FOR THE PATIENT UNDER THE HYPOTHESIS OF A COMMON LOGISTIC MODEL FOR ALL PATIENTS

Patient	Number of Courses	Number of Toxic Courses (O_i)	Expected Number of Toxic Courses (E_i)	Variance of O_i (V_i)
1	2	2	1.52	0.3631
2	1	1	0.69	0.2141
3	6	2	1.62	0.3374
4	4	4	2.40	0.5312
5	6	2	2.82	1.2274
6	0	0	1.26	0.7591
7	6	1	1.31	0.8046
8	5	1	0.56	0.2944
9	2	2	1.72	0.2365
10	5	0	0.59	0.5055
11	5	1	1.19	0.7442
12	3	1	1.86	0.6772
13	1	1	0.89	0.0977
14	4	1	0.57	0.4125

$$T' = \sum_{1}^{14} \frac{(O_i - E_i)^2}{V_i} = 12.4787 \ ,$$

where $E_i = \Sigma P$ for the i-th patient

$V_i = \Sigma P(1-P)$ for the i-th patient

P = estimated probability, using MLE of β and common α.

$$P(Z = 1|x) = \frac{1}{1 + e^{-(\alpha_i + \beta x)}} \ ,$$

where α_i = the i-th patient effect.

Then the likelihood ratio test could be used, using the likelihood for common α versus that obtained under this model. We find find, of course, that the estimates of α_i will not converge for any patient with a uniform series of toxic or nontoxic courses, but the maximum likelihood can be approximated, and the following results are obtained (L denotes log likelihood, and the second expression is evaluated after 20 iterations; λ denotes the likelihood ratio statistic):

$$L(\hat{\alpha}, \ \hat{\beta}) = -22.1300$$

$$L(\hat{\alpha}_i, \ \hat{\beta}) = -12.9320$$

$$\lambda = -2[L(\hat{\alpha}, \ \hat{\beta}) - (L(\hat{\alpha}_i, \ \hat{\beta}))]$$

$$= 18.3960 \ .$$

Under the null hypothesis of no differences among patients (α_i all equal), λ will have an asymptotic χ^2 distribution, with 13 degrees of freedom; the tail area for $\lambda = 18.4$ is approximately 15%. Thus this test does suggest the existence of patient differences though the P value, at that level, is certainly not persuasive.

The observations and the maximum likelihood estimates (MLE's) of the probabilities, under the null hypothesis, are shown in Table 6, for each patient, so that the reader can check the goodness of fit of the null model for the individual patients.

It is interesting to note that the estimate of β is nearly the same under the two models, but the standard error, of course, is quite a bit larger under the broader model. In fact, the hypothesis that $\beta = 0$ is rejected under the first model, but not under the second. Following is a comparison of the MLE's of β under the two models:

TABLE 6. VELBAN DOSE AND TOXICITY, WITH LOGISTIC ESTIMATE OF PROBABILITY OF TOXIC COURSE, ASSUMING NO DIFFERENCES AMONG PATIENTS

Patient	Velban Dose (mg)	Toxicity	Logistic Probability	Patient	Velban Dose (mg)	Toxicity	Logistic Probability
A	0.201	Y	0.72		0.180	N	0.36
	0.207	Y	0.80		0.183	N	0.41
					0.179	Y	0.34
B	0.199	Y	0.69		0.173	N	0.25
C	0.199	Y	0.69	F	0.184	N	0.43
	0.110	N	0.00		0.181	N	0.38
	0.218	Y	0.90		0.183	N	0.41
	0.123	N	0.01		0.143	N	0.04
	0.124	N	0.01		0.089	N	0.00
	0.122	N	0.01				
D	0.202	Y	0.73	G	0.132	N	0.02
	0.204	Y	0.76		0.147	N	0.05
	0.214	Y	0.87		0.177	Y	0.31
	0.144	Y	0.04		0.188	N	0.50
E	0.218	Y	0.90		0.182	N	0.39
	0.191	N	0.55		0.143	N	0.04

TABLE 6 (Continued)

Patient	Velban Dose (mg)	Toxicity	Logistic Probability	Patient	Velban Dose (mg)	Toxicity	Logistic Probability
H	0.189	Y	0.52		0.186	Y	0.47
	0.123	N	0.01		0.156	N	0.09
	0.129	N	0.01		0.156	N	0.09
	0.125	N	0.01		0.156	N	0.09
	0.126	N	0.01	M	0.188	Y	0.50
I	0.220	Y	0.91		0.194	N	0.61
	0.208	Y	0.81		0.203	N	0.75
J	0.158	N	0.10	N	0.217	Y	0.89
	0.159	N	0.11	O	0.181	N	0.38
	0.152	N	0.07		0.127	N	0.01
	0.154	N	0.08		0.156	N	0.09
	0.171	N	0.23		0.156	Y	0.09
L	0.185	N	0.45				

	Null Model	Patient Difference Model
$\hat{\beta}$	72.1	75.6
$SE(\hat{\beta})$	20.5	839.2

On the basis of the results in this section and the preceding
sections, we conclude (1) that the null model is reasonably good,
that is, that patient differences need not be allowed by the
model, and (2) that the velban dose is an important predictor of
toxicity. We propose to use the smooth curve in Figure 3 to predict this toxicity.

Influence And Prognostic Value Of Other Variables

Other variables in Table 1 can be examined for their predictive value, either alone or jointly with the velban dose. The
easiest screening test for the variables, individually, is to
compare levels of each variable for toxic episodes versus levels
of the variable for nontoxic episodes. In Table 7 such analyses
are presented for days since last course, Karnofsky status, platinum dose and bleomycin dose. The only variable among the four
that shows a strongly significant difference between the toxic
and nontoxic courses is the Karnofsky status (P < 1%). There is
a difference of about eight days in the average elapsed time
between courses for the two groups of treatment courses, the
toxic courses coming earlier, but the difference is not highly
significant (P $\cong$ 6.5%, one tailed).

The Karnofsky status was investigated further, since it was
felt that this measure of physical status, related to toxicity,
might be used with the velban dose to predict toxicity more
closely. A multiple logistic model was fitted to the data.

$$P(Z = 1|x,y) = \frac{1}{1 + e^{-(\alpha + \beta x + \gamma y)}} \, ,$$

where   x = velban dose in mg/day kg ,

   y = Karnofsky status .

The correlation between Karnofsky status and velban dose was
checked to make certain it was low enough to avoid the conver-
gence problems that attend the use of coincident variables. The
correlation between the two variables was calculated as -.34.

TABLE 7. COMPARISON OF EACH OF FOUR PREDICTOR VARIABLES FOR TOXIC VERSUS NONTOXIC COURSES

Variable	Nontoxic Courses (36)		Toxic Courses (19)		Normal Test Value
	Mean	Standard Error	Mean	Standard Error	
Karnofsky score	92.50	1.40	83.16	2.17	2.75
Platinum dose	178.28	5.33	177.47	4.36	0.11
Bleomycin	67%[a]	8%	84%[a]	9%	1.41
Days since last course[b]	38.12	3.34	30.9	2.05	1.84

[a]Proportion of courses with bleomycin given.

[b]Omitting first course for each patient.

145

This is satisfactorily low; note the scatterdiagram (Figure 4).
The correlation is significantly different from zero (P $\cong$ 2% by
large sample bivariate normal approximation), indicating that the
velban dose tended to be <u>larger</u> for the <u>sicker</u> patients, a somewhat surprising result (the Karnofsky score, calculated velban
dose, and toxicity outcome are tabled by patient and course in
Table 8), but this lent further hope for the usefulness of the
Karnofsky status as a prognostic factor <u>additional</u> to the velban
dose.

When the joint or multiple logistic model was fit, the results
were as tabulated in Table 9.   Results obtained earlier for the
fit to the velban data, without use of the Karnofsky values, are
given for comparison.   Note that both variables have statistically significant regression coefficients.   However, it can also
be seen that, despite the fact that the increase in the likelihood for Karnofsky status, given the velban dose, is

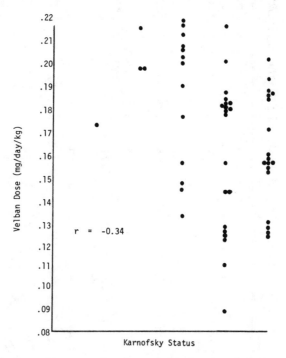

Figure 4.   Scatterdiagram showing the relationship between
Karnofsky status and velban dose (mg/day kg) for 14 patients and
55 treatment courses.

TABLE 8. KARNOFSKY SCORE, VELBAN DOSE (MG/DAY KG) AND GI TOXICITY FOR 14 PATIENTS AND 55 COURSES OF THERAPY

Patient	Karnofsky Status	Velban Dose	Toxicity	Patient	Karnofsky Status	Velban Dose	Toxicity
A	80	0.201	Y		90	0.180	N
	80	0.207	Y		90	0.183	N
B	70	0.199	Y		90	0.179	Y
C	70	0.199	Y		60	0.173	N
	90	0.110	N	F	90	0.184	N
	90	0.218	Y		90	0.181	N
	90	0.123	N		90	0.183	N
	90	0.124	N		90	0.143	N
	90	0.122	N		90	0.089	N
D	90	0.202	Y	G	80	0.132	N
	80	0.204	Y		80	0.147	N
	80	0.214	Y		80	0.177	Y
	80	0.144	Y		90	0.188	N
E	80	0.218	Y		90	0.182	N
	80	0.191	N				

147

TABLE 8 (Continued)

Patient	Karnofsky Status	Velban Dose	Toxicity	Patient	Karnofsky Status	Velban Dose	Toxicity
H	100	0.189	Y		100	0.186	Y
	100	0.123	N		100	0.156	N
	100	0.129	N		100	0.156	N
	100	0.125	N		100	0.156	N
	100	0.126	N	M	100	0.188	Y
I	80	0.220	Y		100	0.194	N
	80	0.208	Y		100	0.203	N
J	100	0.158	N	N	70	0.217	Y
	100	0.159	N	O	90	0.181	N
	100	0.152	N		90	0.127	N
	100	0.154	N		90	0.156	N
	100	0.171	N		80	0.156	Y
L	100	0.185	N				

148

TABLE 9. MULTIPLE LOGISTIC RESULTS FOR VELBAN IN MG/DAY KG (β)
AND KARNOFSKY (γ)

	Velban Alone	Velban and Karnofsky
Velban		
$\hat{\beta}$	72.1	64.4
$SE(\hat{\beta})$	20.5	20.9
Karnofsky		
$\hat{\gamma}$	–	-0.07
$SE(\hat{\gamma})$	–	0.04
SS^a	57.80	39.54

^{a}SS = sum of weighted squared residuals [i.e., T in equation (1)].

statistically significant, comparison with the results for velban
dose alone shows that the regression coefficient for velban does
not change much (64.4 versus 72.1). However, the reduction in
the weighted sum of squared residuals (from 57.80 to 39.54) may
be of some practical importance.

Even though it had seemed slightly more useful to calculate
velban dose relative to the weight of the patient than relative
to his body surface area, it was of some interest to try the suf-
face area method in combination with the Karnofsky status. Since
the surface area would not vary with the temporal variation in
physical status (indeed, it was calculated only once for each
patient), this analysis might be expected to yield two variables
that are less correlated, and thus an analysis that more clearly
separates the prognostic values of dose and physical status. The
results shown in Table 10 were obtained. Note that the results
are very nearly the same as those obtained with velban measured
by kilogram of body weight, namely the regression coefficient for
velban being much the same with or without conditioning on
Karnofsky status, the regression coefficient of Karnofsky status
showing statistical significance along with that for velban, and
the reduction in sum of squares when Karnofsky status is added
being modest but perhaps of some practical importance. Using
both variables, the sum of squared residuals is very nearly the
same for the two methods of calculating dose, namely 39.54 versus
41.98. However, it should be noted that the Karnofsky regression

TABLE 10. MULTIPLE LOGISTIC RESULTS FOR VELBAN IN MG/DAY m^2 (β), WITH AND WITHOUT KARNOFSKY STATUS (γ)

	Velban Alone	Velban and Karnofsky
Velban		
$\hat{\beta}$	1.20	1.51
SE($\hat{\beta}$)	0.40	0.51
Karnofsky		
$\hat{\gamma}$	–	-0.13
SE($\hat{\gamma}$)	–	0.04
SS^a	51.45	41.98

^{a}SS = sum of weighted squared residuals [i.e., T in equation (1)].

coefficient is much stronger (-0.07 versus -0.13) when the dose is measured per square meter and the estimate is statistically significant at a lower level (P = .001 versus P = .08). This, together with custom, would suggest that the results per surface area be used for prediction, rather than those obtained for velban dose per kilogram body weight. The logistic function is plotted in Figure 5. It can be seen that the dependence on both velban dose and Karnofsky status is quite remarkable.

More could have been done with these data along the lines of searching for multiple regression predictors of toxicity. However, time limitations and lack of encouraging results in the univariate analyses of the other variables (recall Table 7) led to termination of the analysis at this point.

CONCLUSION

It was concluded that:

1. The velban dose is a strong predictor of toxicity and the relationship seems to be relatively constant across patients.
2. Karnofsky status is of considerable additional value as a predictor of toxicity.

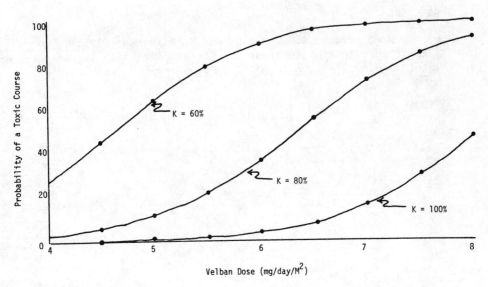

Figure 5. Probability of toxic course estimated from fitted logistic curve using velban dose (mg/day m^2) and Karnofsky status as predictors.

3. There is little or no advantage in calculating velban per kilogram of body weight, rather than per square meter, as a predictor of toxicity.

4. None of the other variables analyzed--cis-platinum dose, bleomycin dose, days since last course--have value as predictors of GI toxicity.[1]

Based on these conclusions, it was recommended that Figure 5 be used in predicting toxicity, using the estimated multiple logistic regression function relating toxicity to velban dose (in mg/day m^2) and to Karnofsky status.

[1]This conclusion has been further strengthened by additional analyses since this report was written. In particular, it was found that the number of days since last course of therapy did not add anything of prognostic value, given the velban dose and Karnofsky status.

REFERENCE

Ayer, M., H. D. Brunk, G. M. Ewing, and E. Silverman (1955). An empirical distribution function for sampling with incomplete information. Ann. Math. Stat., 26, 641-647.

WHICH OF TWO MEASUREMENTS IS BETTER?

BRADLEY EFRON

Medical Problem. Two measurement techniques, one difficult to perform, the other easy, are available for measuring lymphocyte levels in human subjects. Can the easier technique be substituted for the difficult one without loss of important information?

Medical Investigator. Golrukh Vakil, Children's Hospital, Stanford.

Statistical Procedures. Principal components; components of variance; analysis of variance.

MEDICAL BACKGROUND

In this experiment, blood taken from normal human subjects was treated with varying doses of PHA (Phytohaemmaglutnin), which stimulates the production of lymphocytes. The following day, a lymphocyte count was performed using a radioactive labeling technique. Two methods of counting, "regular" and "micro," are available, the second of which is considerably easier to perform than the first. For day-to-day diagnostic use on possibly sick patients it would be convenient to use micro instead of regular. The question is: Can this be done without significant loss of medical information?

THE DATA

Tables 1 and 2 present the count data (number of scintillation counts recorded in a fixed amount of time) for regular and micro, respectively. The asterisks (and subject numbers) show the 17 subjects whose blood was tested on both methods. In addition, 19 subjects were tested on regular only, 23 subjects on micro only. Each subject's blood was tested up to 10 times on each method, at different levels of PHA concentration, ranging from zero to a high of 20 µg. (Subject 24, for example, had 20 separate blood counts performed.) Most of the subjects were not measured on all 10 levels on either method, as shown in Tables 1 and 2.

PRELIMINARY STATISTICAL ANALYSIS

Let x_{ij} be the <u>logarithm</u> of the count (either regular or micro, as will be apparent from the context) obtained on subject i at PHA concentration j. The index j takes on values $1, 2, 3, \ldots,$ 10, corresponding to the 10 possible PHA concentrations. By working with logarithms we stabilize the variance of the measurements, a fact verified by the results of the main statistical analysis.

The simplest model we might hope to fit to the data of Table 1 or 2 is

$$x_{ij} = \alpha_i + \beta_j + e_{ij} , \tag{1}$$

where α_i is the main effect for subject i,

β_j is the main effect for concentration j, and

e_{ij} is the error term.

Of course, the values of (i,j) contemplated in (1) are only those for which data are available. In model (1) it is reasonable to take the β_j to be fixed effects, the α_i to be independent random effects with mean zero and variance σ_α^2,

$$\alpha_i \overset{ind}{\sim} (0, \sigma_\alpha^2) , \tag{2}$$

and the e_{ij} to be independent with mean zero and variance σ_e^2,

TABLE 1. SCINTILLATION COUNT DATA FOR THE REGULAR METHOD. COUNTS WERE OBTAINED FOR 36 SUBJECTS, AT UP TO 10 LEVELS OF PHA CONCENTRATION EACH. THE ASTERISKS INDICATE THE 17 SUBJECTS ALSO TESTED BY THE MICRO METHOD

Subject Number and Date Tested	PHA Concentration									
	j=1 0	j=2 0.1	j=3 0.25	j=4 0.5	j=5 1.0	j=6 2.5	j=7 5.0	j=8 7.5	j=9 10.0	j=10 20.0
1 (1/29/75)	11069	11495	11847	14532	14204	30457	48818	48813	44933	
2 (1/29/75)	12667	14365	16183	16239	12686	27785	45996	46336	50770	
3 (1/29/75)	16422	16465	16247	16730	18782	30362	41127	44222	47307	
4 (2/3/75)	10803	10222	11018	11288	9491	20611	26779	32777	38006	
5 (2/3/75)	6966	6429	6447	7528	7831	16332	30635	35139	36645	
6 (2/6/75)	8578	8048	7982	8095	14574	23884	32782	32203	35415	
7 (2/6/75)	6696	6074	6880	7009	9951	26545	25537	26188	27876	
8 (2/7/75)	11125			11700	16815	32860	38376	39667	28952	25790
9 (2/19/75)	3964		6546		9013	18034	23322	23410	23960	18667
10 (2/21/75)	7466	7552	7370	7577	9043	5939	17001	17393	21116	19558
11 (2/28/75)	5241					5570	14653	13744	18990	20195
12 (4/24/75)	7260	8302		24869	30361	31904	29594	25185	24653	21260
13 (4/30/75)	9938	11594	19384	27223	28660	21910	17394	17890	23195	18771
14 (5/2/75)	9968		16987			25977	22875			
15 (5/5/75)	10997					33392	28235	31598	29507	

TABLE 1 (Continued)

Subject Number and Date Tested	PHA Concentration									
	j=1 0	j=2 0.1	j=3 0.25	j=4 0.5	j=5 1.0	j=6 2.5	j=7 5.0	j=8 7.5	j=9 10.0	j=10 20.0
*16 (5/6/75)	4379		13287	12962	12457	18375	16147	12158		12378
*17 (5/7/75)	3730				8296	13022	10982	10124	9981	
18 (5/10/75)	3663				15286	12978	13825	14259		
*19 (5/13/75)	7592	7312	7226	13710	24853	29593	26062	25055	24086	
*20 (5/14/75)	17702	19013	26153	47790	45259	43097	41731	35776	46013	34204
21 (5/21/75)	7494			19651	37865	35689	32401	31445		
22 (5/23/75)	4166						22079	25751	21706	
*23 (5/28/75)	8287	12866	28193	26910	27003	32441	36780		30149	21850
*24 (5/30/75)	9093	12609	26895	24355	31000	30844	29026	23907	26342	11227
*25 (6/3/75)	8094	10719	33062	32967	34804	32200	31383	27686	28377	19608
*26 (6/6/75)	3154		5168		11953					6621
28 (7/3/75)	5190					41010	27661		25767	21624
*32 (7/25/75)	8903					27640	26007			12385
*36 (8/11/75)	7013				28800	24911	20661	24550	18259	17140
*38 (8/14/75)	10172	8796	9411	26953	35518	36896	27732	27361	27174	22236
*39 (8/18/75)	9774					25116	30855		8648	24414

156

TABLE 1 (Continued)

Subject Number and Date Tested	PHA Concentration									
	j=1	j=2	j=3	j=4	j=5	j=6	j=7	j=8	j=9	j=10
	0	0.1	0.25	0.5	1.0	2.5	5.0	7.5	10.0	20.0
*40 (8/21/75)	3993	6457	3613	14430	15228	16012	11045	12602	13144	9816
*41 (8/25/75)	4044	5164	4707	16142	16921	19695	14119	14165	11552	13148
*48 (12/8/75)	5679	7052	13281	13921	15767	7506	6801	4709	6836	5367
*55 (1/13/76)	9877			29224	31527	31425	26696		17386	
*59 (1/19/76)	9723				23751	20415	11213			

157

TABLE 2. SCINTILLATION COUNT DATA FOR THE MICRO METHOD. COUNTS WERE OBTAINED FOR 40 SUBJECTS, AT UP TO 10 LEVELS OF PHA CONCENTRATION EACH. THE ASTERISKS INDICATE THE 17 SUBJECTS ALSO TESTED BY THE REGULAR METHOD

PHA Concentration

Subject Number and Date Tested	j=1 0	j=2 0.1	j=3 0.25	j=4 0.5	j=5 1.0	j=6 2.5	j=7 5.0	j=8 7.5	j=9 10.0	j=10 20.0
*16 (5/6/75)	7381		16603	20396	29705	30511	20460	15000	18043	17709
*17 (5/7/75)	4945					18130	22653			
*19 (5/13/75)	7150	7150	8671	19676		32367	31081	29062	27519	42140
*20 (5/14/75)	21551		22024	40194	20093	58315	52024	48947	49957	13432
*23 (5/28/75)	10376	13485	15958	26623	35411		34594	33939	28973	26627
*24 (5/30/75)	8920	12128	28386	31584	28141	34315	22032	20691	25906	14974
*25 (6/3/75)	9895	15912	36133	39547		40577	34916	24737	24511	
*26 (6/6/75)	3679						9626			7619
27 (6/11/75)	3949				7951	20511	20319		20955	15789
29 (6/27/75)	2179	7278	*8389	17993		22949	15856	18116	12928	10447
30 (7/10/75)	8729			*36824	50935	41889	45080	39628	36754	30781
31 (7/18/75)	7443	8048	12598	30343	32427	26247	22155	21095	20230	11863
*32 (7/25/75)	9006	8660	23305	33043	40160	39563	31777	28702	33122	25262
33 (8/1/75)	6679	11849	8594	37401	36418	35922	29238	28617	22780	15831
34 (8/1/75)	7884	8506	8700	30622	30532	32033	28130	24897	23368	18036

158

TABLE 2 (Continued)

Subject Number and Date Tested		PHA Concentration									
	j=1 0	j=2 0.1	j=3 0.25	j=4 0.5	j=5 1.0	j=6 2.5	j=7 5.0	j=8 7.5	j=9 10.0	j=10 20.0
35 (8/6/75)	9224				17166		26548			26536
*36 (8/11/75)	7669	12302	9021	34780	30095	33208	29585	29208	24543	23076
37 (8/13/75)	8413	10177	10777	33277	34188	36069	36038	30810	25869	22435
*38 (8/14/75)	8441	18480	12296	46532	48233	50454	41097	33428	32104	22656
*39 (8/18/75)	10232		10132	36849	36876	37042	34389		28876	24414
*40 (8/21/75)	5652	9019	5983	24493	24527	25431	20121	15887	19502	14250
*41 (8/25/75)	5780	9540	7229	25648	20864	22597	23384	19359	18637	15723
42 (8/4/75)	722						2353			
43 (9/8/75)	2246				14397	11195	11781			
44 (9/18/75)	7290	7436	20834	19017	24733		33692	32712	32776	28252
45 (9/22/75)	3268	3339	9217	12327	11107	14281	13537	13101	11020	
46 (11/21/75)	4575	10046	17331	19577	16864	14193	11168	9053	11992	9490
47 (12/3/75)	5373	6621	8217	11713	16513	16953	19504	16006	18824	14743
*48 (12/8/75)	5679	4951	12245	15804	16406	15224	5870	4709	6836	5367
49 (12/18/75)	5475	7596	15560	27748	30328	24064	19931	14941	17554	10797
50 (12/29/75)	5277	4895	8207	13248	16253	14344	12345	12737	11153	7077

159

TABLE 2 (Continued)

Subject Number and Date Tested	PHA Concentration									
	$j=1$ 0	$j=2$ 0.1	$j=3$ 0.25	$j=4$ 0.5	$j=5$ 1.0	$j=6$ 2.5	$j=7$ 5.0	$j=8$ 7.5	$j=9$ 10.0	$j=10$ 20.0
51 (12/30/75)	5792	5183	7387	13162	15649	15717	14923	14695	13866	8156
*52 (1/5/76)	15607	18256	25075	33966	44745	41230	37115	34769	26238	13420
53 (1/6/76)	15122		22985	34743	51197	40740	39448		26746	23166
54 (1/7/76)	13828	12991	20373	40565	40578	32106	32432	27411	24052	16829
*55 (1/13/76)	9863	10705	23493	35277	39303	39671	34774	22016	25416	19992
56 (1/15/76)	11431			32042	31018	25565	29357			
57 (1/16/76)	5015			14714	22960	18266	14996		17708	
58 (1/19/76)	20709			55711	56229	43589	26952		20451	
*59 (1/22/76)	11527	6314	22031	25382	24053	17035	11093	11676	7551	5074

160

$$e_{ij} \overset{\text{ind}}{\sim} (0, \sigma_e^2) \; . \tag{3}$$

(For identifiability purposes, the constraint $\Sigma \alpha_i = 0$ was imposed on model (1), which does not exactly agree with assumption (2), but causes no trouble in the subsequent analysis.)

A persuasive definition of a <u>good</u> measurement technique is one whose response changes a lot, and in a dependable way, when we change the quantity being measured (in this case, lymphocyte level), and not very much when we change the experimental unit upon which the measurement is performed. In terms of model (1), micro would be as good, or better, than regular if

$$\beta_j(\text{micro}) = \beta_j(\text{regular}) \qquad j = 1,2,\ldots,10 \tag{4}$$

and if

$$\sigma_\alpha^2(\text{micro}) \leq \sigma_\alpha^2(\text{regular}) \; ,$$

$$\sigma_e^2(\text{micro}) \leq \sigma_e^2(\text{regular}) \; . \tag{5}$$

The situation would be even more favorable to micro vis-à-vis regular if (5) held and the β_j(micro) varied more rapidly than did the β_j(regular), as the lymphocyte level varied. We cannot actually measure the true lymphocyte level, but sometimes this more favorable version of (4) can be inferred from plots of estimated β_j values versus j, as in the main statistical analysis.

A preliminary statistical analysis was run using model (1). The parameters α_i and β_j were fit by least squares, separately for the regular and the micro data. The fitted values $\hat{\beta}_j$, $\hat{\sigma}_\alpha^2$, $\hat{\sigma}_e^2$ did indeed satisfy (5) and the more favorable version of (4), indicating superiority for the micro process. However, a principal components analysis of the residuals from model (1), which will now be described, disclosed the inadequacy of this model.

Define

$$r_{ij} = x_{ij} - [\hat{\alpha}_i + \hat{\beta}_j] \; , \tag{6}$$

the residual from the least squares fit to model (1). Also define the 10×10 matrix C whose (a,b)-th element is

$$c_{ab} = \sum_i \frac{r_{ia} r_{ib}}{n_{ab}} , \qquad (7)$$

the sum in (7) being over those values of i having both r_{ia} and r_{ib} defined, and n_{ab} being the number of such i values.[1] If there were no missing values, C would be (almost) the usual estimate of the residual covariance matrix. There is some danger in using (7), the "pairwise present" definition of C (see Heiberger, 1977), but the obvious alternative, maximum likelihood estimation, was not easily available.

A principal components analysis was run on the matrix C obtained from each set of data. For regular, the first principal component accounted for 59% of the total residual variation from model (1), i.e., 59% of $\Sigma_{i,j} r_{ij}^2$. The second principal component accounted for only 19% of the residual variation, the third 12%, the fourth 6%. For micro, the first principal component accounted for 50% of the residual variation from model (1), the second component 20%, the third 11%, the fourth 9%.

Without attempting a formal significance test, which would be difficult even under normal theory given the missing (i,j) values in model (1), this pattern suggests expanding model (1) to include the first principal component, say

$$x_{ij} = \alpha_i + \beta_j + \gamma_i z_j + e_{ij} . \qquad (8)$$

Here z_j is the j-th coordinate of the first principal component, and γ_i is another random effect connected with the different subjects. For example, if $z_j = j - 5.5$ (which it does not), then γ_i would be a slope effect measuring how quickly subject i's responses changed with increasing PHA concentration.

Table 3 shows the first principal component vector for regular and micro. These agree rather nicely. In order to work with the same model (8) for both methods, a stylized z vector was actually used,

[1]As before, we are doing separate analyses for the micro and regular data, the only connection being that some of the subjects show up in both data sets.

TABLE 3. THE FIRST PRINCIPAL COMPONENT (EIGENVECTOR) OF C FOR
REGULAR AND FOR MICRO, AND THE STYLIZED VERSION USED IN THE MAIN
STATISTICAL ANALYSIS

Vector	Component of C									
Used	1	2	3	4	5	6	7	8	9	10
Regular	.05	−.06	−.38	−.48	−.44	−.02	.27	.40	.37	.25
Micro	−.29	−.09	−.63	−.14	−.23	.07	.29	.33	.32	.38
Stylized	0	0	−1/3	−1/3	−1/3	0	1/4	1/4	1/4	1/4

$$z_j = \begin{cases} -\dfrac{1}{3} & j = 3, 4, 5 , \\[2mm] \dfrac{1}{4} & j = 7, 8, 9, 10 , \\[2mm] 0 & j = 1, 2, 6 . \end{cases} \tag{9}$$

As a compromise, (9) fits the regular vector more closely than
the micro vector, but specification errors of this type are not
likely to be disasterous, and in any case should err in the conservative direction, since we have a vested interest in demonstrating the superiority of micro. Notice that the z vector (9)
is orthogonal to $(1,1,1,\ldots,1)$, so that α_i and γ_i are measuring
orthogonal linear combinations of subject i's response, at least
for those subjects having all ten measurements.

MAIN STATISTICAL ANALYSIS

   Model (8), (9) was fit by least squares to the regular data of
Table 1, and, separately, to the micro date of Table 2. Table 4
displays the results. (Note: the constraints $\Sigma\alpha_i = \Sigma\gamma_i = 0$ were
imposed to guarantee identifiability.)
   A close look at Table 4 shows that the micro measurement
method is superior to, or at least as good as, the regular measurement method, in the sense of (4), (5). This is the main conclusion of the statistical analysis. Figure 1 graphs the $\hat{\beta}_j$ for
the two methods. The micro values rise more quickly and then

TABLE 4.   LEAST SQUARE ESTIMATES OF PARAMETERS IN MODELS (8), (9), FOR REGULAR AND MICRO DATA.   THE ESTIMATES $\hat{\sigma}_\alpha$, $\hat{\sigma}_\gamma$ ARE OBTAINED BY STANDARD METHODS OF MOMENTS CALCULATIONS, AS IN SCHEFFE (1959), SECTION (7.2).   SUBJECT 42, WHO WAS ONLY MEASURED AT TWO PHA CONCENTRATIONS, WAS NOT INCLUDED IN THE ESTIMATE OF $\hat{\sigma}_\alpha$ (MICRO)

Quantity	Regular	Micro
Measurement error $\hat{\sigma}_e$	0.221	0.238
α component of error $\hat{\sigma}_\alpha$	0.408	0.404
γ component of error $\hat{\sigma}_\gamma$	1.190	0.734
Main effects due to changing PHA concentration		
$\hat{\beta}_1$	8.91	8.84
$\hat{\beta}_2$	9.05	9.04
$\hat{\beta}_3$	9.34	9.38
$\hat{\beta}_4$	9.63	10.02
$\hat{\beta}_5$	9.78	10.09
$\hat{\beta}_6$	9.99	10.14
$\hat{\beta}_7$	10.04	9.99
$\hat{\beta}_8$	10.00	9.88
$\hat{\beta}_9$	10.00	9.85
$\hat{\beta}_{10}$	9.81	9.56

fall off more quickly. Since it is expected, from past experience, that the stimulation effect of PHA diminishes at large concentrations, this seems to confirm that the micro technique is responding more quickly to changes in the actual blood lymphocyte level of the subjects.

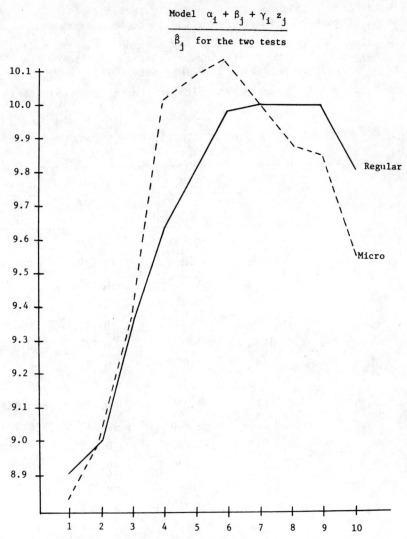

$$\text{Model } \alpha_i + \beta_j + \gamma_i z_j$$

$$\hat{\beta}_j \text{ for the two tests}$$

Figure 1. Fitted parameters $\hat{\beta}_j$ for the regular and micro
measurement data.

Tables 5 and 6 show the fitted values $\hat{\alpha}_i$, $\hat{\gamma}_i$ for the 17 common subjects, and for the noncommon subjects, respectively. The pairs $(\hat{\alpha}_i(\text{regular}), \hat{\alpha}_i(\text{micro}))$ for the 17 common subjects have correlation 0.91, which reiterates how similarly the two models are behaving in this respect. The correlation coefficient for $(\hat{\gamma}_i(\text{regular}), \hat{\gamma}_i(\text{micro}))$ is a less impressive 0.67. However, missing data made $\hat{\gamma}_i$ highly variable for some of the subjects

TABLE 5.   FITTED VALUES $\hat{\alpha}_i$, $\hat{\gamma}_i$ FOR THE 17 COMMON SUBJECTS

Subject Number	Regular Method		Micro Method	
	$\hat{\alpha}_i$	$\hat{\gamma}_i$	$\hat{\alpha}_i$	$\hat{\gamma}_i$
16	-0.316	-0.561	0.060	-0.330
17	-0.703	-0.096	-0.332	1.469
19	0.044	0.333	0.088	1.024
20	0.772	-0.612	0.759	1.059
23	0.406	-0.553	0.345	0.099
24	0.285	-1.071	0.330	-0.357
25	0.409	-1.061	0.499	-1.007
26	-0.840	-0.729	-0.627	-0.383
32	0.214	-0.708	0.420	-0.093
36	0.127	-0.850	0.259	0.395
38	0.277	-0.263	0.529	0.088
39	0.212	-2.120	0.381	0.369
40	-0.498	-0.263	-0.097	0.215
41	-0.397	-0.313	-0.053	0.345
48	-0.676	-2.152	-0.667	-1.631
55	0.270	-1.108	0.389	-0.411
59	-0.088	-1.633	-0.258	-1.775

TABLE 6.   FITTED VALUES OF $\hat{\alpha}_i$, $\hat{\gamma}_i$ FOR THE 19 SUBJECTS TESTED ONLY BY THE REGULAR METHOD, AND FOR THE 23 SUBJECTS TREATED ONLY BY THE MICRO METHOD

	Regular Method			Micro Method	
Subject Number	$\hat{\alpha}_i$	$\hat{\gamma}_i$	Subject Number	$\hat{\alpha}_i$	$\hat{\gamma}_i$
1	0.381	1.421	27	-0.425	1.937
2	0.442	1.256	29	-0.342	0.061
3	0.508	0.905	30	0.597	0.213
4	0.106	1.195	31	0.070	-0.392
5	-0.108	1.898	33	0.222	0.065
6	0.051	1.325	34	0.155	0.263
7	-0.137	1.375	35	0.141	1.242
8	0.238	0.968	37	0.311	0.363
9	-0.281	1.169	42	-2.256	0.105
10	-0.372	0.817	43	-0.773	-0.275
11	-0.852	2.444	44	0.281	0.688
12	0.246	-0.552	45	-0.586	0.072
13	0.180	-1.081	46	-0.323	-0.926
14	0.204	-0.669	47	-0.257	0.671
15	0.414	-0.501	49	-0.039	-0.718
18	-0.446	-0.470	50	-0.490	-0.223
21	0.363	-0.294	51	-0.411	0.160
22	-0.573	2.436	52	0.523	-0.434
28	0.141	0.121	53	0.560	-0.238
			54	0.581	-1.206
			56	0.416	-0.461
			57	0.281	-0.046
			58	-0.260	0.003

(a fact that should be kept in mind in comparing $\hat{\sigma}_\gamma$(regular) with $\hat{\sigma}_\gamma$(micro) in Table 4). A disturbing feature of Table 5 is that 16 of the 17 $\hat{\gamma}_i$ values are negative. Remembering that the sum of all 36 = 17 + 19 $\hat{\gamma}_i$ values for regular equals zero, this suggests some unexplained difference in laboratory methods used for the 17 common subjects.

A residual analysis was carried out on the residuals from model (8), (9),

$$r_{ij} = x_{ij} - [\hat{\alpha}_i + \hat{\beta}_j + \hat{\gamma}_i z_j] . \tag{10}$$

Table 7 shows the sample standard deviations at each PHA concentration, $\hat{\sigma}_j = \sqrt{c_{jj}}$, c_{ab} as defined in (7). These are reasonably similar for the 2 methods and for the 10 different values of j. For both tests, the largest value of $\hat{\sigma}_j$ occurs at j = 3. This probably is due to the z_j values changing from 0 to -1/3 at j = 3. A smoother choice of the z vector would alleviate this.

TABLE 7. SAMPLE STANDARD DEVIATIONS OF THE RESIDUALS FROM MODEL (8), (9), BY METHOD AND PHA CONCENTRATION

	Standard Deviation	
j	Regular Method	Micro Method
1	0.230	0.272
2	0.194	0.210
3	0.306	0.290
4	0.156	0.185
5	0.192	0.185
6	0.255	0.119
7	0.157	0.111
8	0.110	0.144
9	0.153	0.094
10	0.164	0.187

No other gross deficiencies of model (8), (9) were apparent.

For the 17 common subjects it is possible to compute correlation coefficients between the residual pairs (r_{ij}(regular), r_{ij} (micro)) for each value of j. These are displayed in Table 8. There are several large correlations, which can be interpreted simply as indicating a close relationship between the two measurement methods. Less optimistically, it might also indicate a deficiency in model (8), (9); correlations near zero could have been interpreted as showing the errors e_{ij} to be inherent in the separate measurement processes themselves, and hence not explainable by the addition of further functions of i to model (8).

DISCUSSION

The characterization of a good measurement method as one that varies when it should and does not when it should not is hard to argue with. However, some forms of "varying when it should not" are less harmful than others. It is clear that a large value of σ_e can only be detrimental. A large value of σ_α, however, may be irrelevant in certain contexts. Suppose that the purpose of the measurement process is to classify a new subject into categories "sick" or "well" on the basis of a weighted linear combination of the log measurements $x_1, x_2, \ldots, x_{10}$, say $\sum_{j=1}^{10} w_j x_j$. If we believe that the overall level of the measurements is greatly affected by factors unrelated to health, such as inherent variability of subjects, changes in lab techniques, and so on, it may make sense to take $\sum_{j=1}^{10} w_j = 0$, in which case the α_i term in a model like (8) is nullified. (As a matter of fact, the linear combination $x_7 - x_1$ is the one which actually has been used for classification purposes.)

A handy but worrisome feature of the analysis used here to compare the two measurement methods is that it does not require any subject to be tested on both methods! The comparison is built into the model (8), (9) and criterion (4), (5). Nevertheless it is reassuring to see, as in Table 5, that the two methods seem to give similar results for the same subject. Otherwise we might worry that the methods were measuring basically different quantities, no matter how similar the estimated parameters β_j, σ_α, σ_γ, σ_e were. Such reassurance can only come from subjects tested by both methods.

TABLE 8.   CORRELATIONS BETWEEN THE RESIDUALS (r_{ij}(REGULAR), r_{ij} (MICRO)) FROM MODEL (8), (9); n INDICATES THE NUMBER OF SUBJECTS HAVING BOTH MEASUREMENTS FOR THAT VALUE OF j

j	1	2	3	4	5	6	7	8	9	10
Correlation	.72	.15	.86	.82	.52	−.03	.08	.78	.48	−.28
n	17	8	10	11	11	15	16	10	12	12

REFERENCES

Heiberger, R. M. (1977).   Regression with the pairwise-present covariance matrix: a dangerous practice.   Technical Report No. 19, Department of Statistics, University of Pennsylvania.

Scheffé, H. (1959).   The Analysis of Variance.   New York: Wiley.

DETERMINING AN AVERAGE SLOPE

JOHN HYDE

Medical Problem. Measure the efficiency with which perfused kidneys use oxygen to reabsorb sodium from urine.

Medical Investigator. Michael Weiner, Stanford University.

Statistical Procedures. Linear regression; analysis of variance; weighting; variance component estimation; jackknife; bias correction.

MEDICAL BACKGROUND

One way to learn about the function of the kidney is to study the rates at which it produces and consumes different substances. An important quantity is the rate at which oxygen is consumed, since this is considered to be a measure of how hard the kidney is working. Another item of interest is the rate at which the kidney reabsorbs ionic sodium from the urine. This activity, known as sodium pumping, requires energy. Thus there should be a direct relationship between the intensity of sodium pumping and the rate of oxygen consumption.

A different investigator had experimented on kidneys in live dogs and determined a value of 29 Eq Na/mole O_2 (equivalents of sodium per mole of oxygen) with a standard error of 3 Eq Na/mole O_2. Our medical investigator had made his observations on rat kidneys that had been perfused, that is, they had been removed from the animals and maintained with fluids. This was a more convenient experimental arrangement, and our investigator was interested, among other things, in seeing if the sodium pumping efficiency was comparable to that obtained previously.

Our investigator also had some questions about the way data from other experiments had been analyzed. In these experiments one or two data points were obtained from each animal. The data were combined and used to estimate the regression of oxygen on sodium. The intercept was interpreted as the basal rate of oxygen consumption, and the reciprocal of the slope was used as the estimate of the pumping efficiency. Our investigator had computed separate slopes for each kidney, but he was not sure how they should be combined and how to compute standard errors.

THE DATA

There are two data sets, but they have a number of features in common. In both, kidneys were maintained by perfusion. The amount of oxygen consumed and the amount of sodium reabsorbed were measured indirectly over 10-minute intervals. The kidneys were given time to stabilize in the test environment, and observations were made before apparent deterioration had set in. Some kidneys were disqualified on the basis of an objective measure of stability and health of the kidney.

The low pressure data were obtained by lowering pressure in order to perturb the kidney equilibrium point and thus provide a range of values for regression. There were 10 kidneys. Five observations were made on each kidney, except that the second contributed only four data points. The data are given in Appendix 1. The first column is the rate of sodium reabsorption in μEq/min, multiplied by 100. The second column shows the rate of oxygen consumption. This is given in μmoles/min, multiplied by 1000.

The ouabain data were obtained by changing the kidney equilibria with the drug ouabain. There were 17 kidneys. The first eight were measured by one experimenter, Maureen, and each kidney provided six data points. The data on the last nine kidneys were contributed by another experimenter, George. These kidneys produced four data points apiece. All 17 kidneys were dried and weighed after the experiment, and their weights were recorded, although they were not used in the analyses presented here. The data are given in Appendix 2. The first column is the dried weight in milligrams, multiplied by 10. The second column shows sodium reabsorption in μEq/min, multiplied by 100. The last column is oxygen consumption in μmoles/min, multiplied by 1000.

STATISTICAL PROLOGUE

Both sodium and oxygen were measured indirectly, and the intervening steps and instruments introduced measurement error. Although he did not know the sizes of these errors, the investigator felt that the measurement of oxygen was much less accurate than the measurement of sodium. Thus it is more appropriate to treat sodium as the independent variable. If sodium does have appreciable error relative to oxygen, then the estimator of the slope will be biased toward zero. [Kendall and Stuart (1961) have an extensive discussion of this problem in Chapter 29.] Investigations are being planned to estimate the measurement error variances.

The quantity of interest, however, is the reciprocal of the regression slope. In general, the reciprocal of an unbiased estimator is biased. Using the first few terms of a Taylor series expansion of x^{-1} around the true parameter value, one can usually approximate the bias adequately in terms of the estimate and its standard error.

The medical investigator was quite correct in his decision to compute separate slopes for the kidneys. If all of the date were combined, differences in basal rates could cause the sodium-oxygen relationship within kidneys to be seriously confounded with the sodium—oxygen relationship across kidneys. For purposes of comparison, we did fit a single regression line to the combined data.

LOW PRESSURE EXPERIMENT

The data from this experiment are described above and presented in Appendix 1.

Let the index i correspond to the kidney number, and let j index observations within a kidney. Let Y stand for oxygen consumption, and X, sodium reabsorption. The model we considered was

$$Y_{ij} = \alpha_i + \beta_i X_{ij} + \varepsilon_{ij}, \qquad j = 1, \ldots, N_i,$$

$$i = 1, \ldots, 10,$$

where N_i is the number of observations on kidney i, and the ε_{ij}'s are independent errors with mean zero and variance σ_e^2. The values of the least squares estimators of β_i, denoted by b_i, are given in Table 1. The variance of b_i was estimated by $\hat{\sigma}_e^2/C_{xx}^{(i)}$, where $\hat{\sigma}_e^2$ is the mean residual sum of squares, and $C_{xx}^{(i)} = \sum_{j=1}^{N_i} (X_{ij} - \bar{X}_i)^2$, that is, $C_{xx}^{(i)}$ is the corrected sum of squares for sodium in kidney i. The values of $C_{xx}^{(i)}$ are also given in Table 1.

An analysis of variance table is shown in Table 2. A test of the hypothesis of equal slopes showed that there was little evidence that the kidneys had different slopes, but the common slope appeared to be nonzero.

The hypothesis of equal error variances in the 10 kidneys was tested by computing the ratio of the largest to smallest variance. The value, 35.7, was less than that given for 10 observations with 3 degrees of freedom in the 5% table of Pearson and Hartley (1954), p. 179. This test should be liberal, since one of the variances had only 2 degrees of freedom. The variances were pooled.

The least squares estimator of β, denoted by b, is related to the individual slope estimators by

TABLE 1. VALUES OF b_i AND $C_{xx}^{(i)}$ FOR THE LOW PRESSURE EXPERIMENT

Kidney	b_i	$C_{xx}^{(i)}$
1	0.00967	1384
2	0.04784	360
3	0.03134	753
4	0.01928	3153
5	0.01928	3050
6	0.01747	4575
7	0.04817	1570
8	0.01893	4175
9	0.04233	719
10	0.02706	885

TABLE 2. ANOVA TABLE FOR LOW PRESSURE EXPERIMENT

Source	S.S.	d.f.	M.S.	F
Common Slope	10.40	1	10.40	31.3
Separate slopes	2.09	9	0.232	0.7
Residual	9.63	29	0.332	

$$b = \frac{\sum_{i=1}^{10} b_i \, c_{xx}^{(i)}}{\sum_{i=1}^{10} c_{xx}^{(i)}} \, ,$$

that is, b is a weighted average of the individual estimators.
Since the variance of b_i is inversely proportional to $c_{xx}^{(i)}$, the
above weighted average has the smallest variance among all
weighted averages, as we expect from linear model theory. Table
3 shows a scatterplot of the b_i's, together with their weights.
The weights have been normed to sum to 100%.

Thus we have one answer to the investigator's question of how
to combine the separate slopes: use weights given in Table 1
under $c_{xx}^{(i)}$. The variance of b is estimated by $\hat{\sigma}_e^2 / \sum_{i=1}^{10} c_{xx}^{(i)}$.

There are other ways in which slopes might be combined, and it
is instructive to try these methods. Although it will have a
larger variance than the least squares estimator, one might wish
to consider the simple (unweighted) average of the 10 slopes.

TABLE 3. SLOPES AND WEIGHTS FOR LOW PRESSURE EXPERIMENT

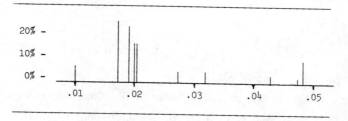

Call this b_{SA}. Estimates of the variance can be obtained in two
ways: the "fixed effects" approach views b_{SA} as a linear combination of the b_i's, so the variance of b_{SA} is a related linear
combination of the separate variances. This is acceptable as
long as the β_i's are considered to be fixed effects. The
"sample" approach computes the variance of the mean of the b_i's
as one would for independent identically distributed observations. Although this seems naive, it produces an unbiased estimator of variance.

One might also apply the jackknife technique. Jackknifing the
least squares estimator b requires that slopes and weights be re-
computed each time a point is removed. One can avoid some of
this complexity by jackknifing the simple average instead, so
that recomputation of weights is unnecessary. [If the β_i's are
considered fixed effects, then such a procedure produces the
unbalanced jackknife estimate of $\Sigma_{i=1}^{10} \beta_i/10$. See Miller (1974).]
The pseudovalues have a particularly simple form. If $b_{i(-j)}$ is
the slope for kidney i after removing observation j, then

$$P_{ij} = b_{SA} + \frac{49-1}{10} (b_i - b_{i(-j)}) \ .$$

The mean slope and its variance are estimated in the usual way by
treating the pseudovalues as a sample. An important property of
the jackknife is its ability to reduce bias, but this is not so
crucial here because both simple and weighted averages are
unbiased. However, the jackknife provides a generally valid
second opinion on the variance, which can be useful for comparisons.

The results for these estimation procedures are given in Table
4. From any estimator $\hat{\beta}$ of β, an estimator of β^{-1} was obtained
by computing

$$(\hat{\beta})^{-1} - \frac{\hat{\sigma}^2}{(\hat{\beta})^3}$$

where $\hat{\sigma}^2$ was the estimated variance of $\hat{\beta}$. This should reduce the
bias introduced by the transformation x^{-1}. Confidence intervals
for β^{-1} were obtained by transforming confidence intervals for β.

TABLE 4. SUMMARY OF ANALYSES OF LOW PRESSURE EXPERIMENT

Method	$\hat{\beta}$	$\hat{\sigma}$	$\hat{\beta}^{-1}$	95% Confidence Set
Combined	0.01381	0.00506	62.7	(42.1, 256.9)
Simple average				
Fixed effects	0.02814	0.00549	34.2	(25.7, 57.5)
Sample	0.02814	0.00434	34.7	(26.5, 53.8)
Weighted average	0.02244	0.00401	43.1	(33.0, 68.6)
Jackknife	0.02188	0.00689	41.1	(28.3, 119.4)

Since the estimators of β have a fairly symmetric distribution while their inverses are skewed, the indirectly obtained confidence sets were deemed more satisfactory. It is for this reason, too, that we decided not to jackknife the seemingly more natural estimators $(\hat{\beta})^{-1}$.

Bias corrections ranged from 0.8 for the sample version of the sample average to 4.5 for the jackknife. The "combined" method is the incorrect analysis which fits only one slope and one intercept. The bias correction was 9.7 for the combined analysis.

OUABAIN EXPERIMENT

These data were obtained by administering the drug ouabain to change the equilibrium levels of kidneys. The data are described earlier and are presented in Appendix 2.

Table 5 gives the values of b_i and $C_{xx}^{(i)}$ for each kidney. Table 6 shows a scatterplot of slopes together with the weights which give the least squares estimator of a common slope. The weights are just the values of $C_{xx}^{(i)}$, normalized to add to 100%. There was a wide range in weights. The ratio of largest to smallest was about 90.

An analysis of variance table is provided in Table 7. The decrease in sum of squares due to including separate slopes was subdivided into the effects due to experimenters, and the effects due to differences within experimenters. There was substantial

TABLE 5.   VALUES OF b_i AND $c_{xx}^{(i)}$ FOR THE OUABAIN EXPERIMENT

Kidney	b_i	$c_{xx}^{(i)}$
1	0.03084	8876
2	0.07907	2337
3	0.04418	687
4	0.05670	531
5	0.03011	3489
6	0.06347	1051
7	0.06092	1470
8	0.06750	102
9	0.05527	1435
10	0.00936	1655
11	0.02475	655
12	0.08028	965
13	0.00845	191
14	0.07218	879
15	0.03818	1347
16	0.05297	1627
17	0.02266	2824

evidence that the common slope was nonzero. In addition, there was evidence that the kidneys had different slopes, and the differences did not appear to be attributable to the fact that there were two experimenters. It is gratifying to note that the residual mean square ($\hat{\sigma}_e^2$) was in close agreement with the value obtained from the low pressure experiment.

Since the degrees of freedom varied, the simple ratio test for equality of variances was not used. Bartlett's test (Snedecor and Cochran, 1967, p. 296) was performed instead. It showed a difference at $p = 0.015$, but the test is sensitive to nonnormality. A close look at the individual variances disclosed

TABLE 6.   SLOPES AND WEIGHTS FOR OUABAIN EXPERIMENT

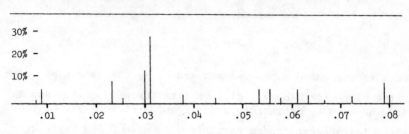

the following: There were 14 kidneys with variances between 0.10
and 0.65.   One value was 0.038, and two kidneys had points sit-
ting nearly on a line and giving variances of about 0.0075.   The
two "inliers" accounted for most of the observed differences.   It
seemed more reasonable to attribute them to good luck rather than
to the possibility that selected kidneys were measured with unus-
ually high precision.   Thus we elected to pool the variances and
gain the stability of greater degrees of freedom.

   The discovery of differences in slopes means there is no common slope to estimate, so $\hat{\beta}$ (a generic estimator) is estimating a
slightly more abstract quantity, namely the mean slope for some
population of kidneys.   Furthermore, estimates of the variance
of $\hat{\beta}$ must be made to account for the sampling variation resulting
from taking a sample of only 17 kidneys from a larger population.
In other words, the slopes should be treated as random effects.
This contributes to the uncertainty in estimating β.

TABLE 7.   ANOVA TABLE FOR OUABAIN DATA

Source	S.S.	d.f.	M.S.	F	P
Common slope	51.230	1	51.230	177.01	<0.001
Separate slopes	11.895	16	0.743	2.57	0.006
For experimenters	0.070	1	0.070	0.24	0.62
Within Maureen's	5.998	7	0.857	2.96	0.012
Within George's	5.827	8	0.728	2.52	0.022
Residual	14.471	50	0.2894		

It is helpful to have a model. If b_i is the estimator of the slope for kidney i, we assume

$$b_i = \beta + \delta_i + \varepsilon_i \; ,$$

where β is the unknown mean slope, and δ_i is the amount by which the true slope for kidney i differs from the mean of the population. The term ε_i represents measurement error. Say that the δ_i's are independent random variable with mean 0 and variance σ_δ^2. Assume that the ε_i's have mean 0 and that they are independent of each other and of the δ_i's. The variances of the ε_i's differ; we are able to estimate them from the regressions we have done:

$$\hat{\sigma}_i^2 \equiv \hat{\text{Var}}(\varepsilon_i) = \hat{\sigma}_e^2 / C_{xx}^{(i)} \; .$$

VARIANCE COMPONENTS

The problem of finding σ_δ^2 is a problem in variance component estimation. The problem at hand is a natural extension of the conventional unbalanced case (see Searle, Chapters 10 and 11 for a useful review). There are a number of ways in which the problem can be attacked, but this is largely a reflection of the fact that no generally accepted means has been found for selecting a "best" procedure.

Consider first the method of moments. If w_i is a sequence of nonnegative weights with $W = \Sigma w_i > 0$, the method of moments produces the unbiased estimator

$$\hat{\sigma}_\delta^2 = \frac{\Sigma w_i (b_i - b_W)^2 - (n-1) \hat{\sigma}_e^2}{W - (\Sigma w_i^2 / W)} \; ,$$

where $\hat{\sigma}_e^2$ is the residual mean sum of squares obtained previously, and b_W is the weighted average of the slopes using weights w_i. The variance of b_W can be estimated by

$$\hat{\sigma}_\delta^2 \cdot \frac{\Sigma w_i^2}{W} + \frac{\hat{\sigma}_e^2}{W} \ .$$

If $w_i \equiv 1$, then b_W is just the simple average b_{SA}. The variance estimator simplifies to the usual variance formula applied formally to the b_i's.

A second choice of weights is $w_i = c_{xx}^{(i)}$, as was done for the low pressure experiment. Call these the raw weights. A rationale for using these weights with the low pressure data was that the variance of b_i was inversely proportional to $c_{xx}^{(i)}$. For the ouabain data $Var(b_i) = \sigma_\delta^2 + \sigma_e^2/c_{xx}^{(i)}$, so one might wish to use weights proportional to $(\sigma_\delta^2 + \sigma_e^2/c_{xx}^{(i)})^{-1}$. Since σ_δ^2 is unknown, one is led to an iterative procedure. Raw weights could be used to obtain a first estimate of σ_δ^2, new weights could be computed, and the process repeated until the estimate stabilizes. Call this a modified weights procedure. The final weights are shown in the scatterplot in Table 8. The ratio of the largest weight to the smallest weight was reduced to about 10. Since approximate weights often perform as well as exact weights, we might expect the first iteration with new weights to be close to the final iteration. The results bore this out.

Since the modified weights are functions of the data, one might worry that the resulting weighted average would be biased. However, the modified weights at each step are functions of $\hat{\sigma}_e^2$ and of a quadratic form in the b_i's, both of which are uncorrelated with the b_i's. Under the assumption of normality, the

TABLE 8.   SLOPES AND MODIFIED WEIGHTS FOR OUABAIN EXPERIMENT

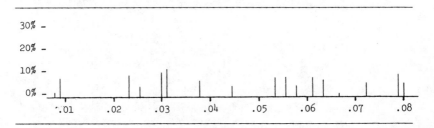

weights will be independent of the b_i's, and in general we can expect the correlations to be small.

The maximum likelihood method can also be used. If the δ_i's and ε_i's are given normal densities, then the likelihood equations are

$$\sum_{i=1}^{17} \frac{(b_i - b_{ML})}{(\hat{\sigma}_\delta^2 + \hat{\sigma}_i^2)} = 0 \, ,$$

$$\sum_{i=1}^{17} \left[\frac{(b_i - b_{ML})^2}{(\hat{\sigma}_\delta^2 + \hat{\sigma}_i^2)^2} - (\hat{\sigma}_\delta^2 + \hat{\sigma}_i^2)^{-1} \right] = 0 \, .$$

The first equation can be solved explicitly to give an expression for b_{ML} which is similar to the modified weighted average. Thus the system can be solved numerically in terms of the single unknown $\hat{\sigma}_\delta^2$. The covariance matrix for b_{ML} and $\hat{\sigma}_\delta^2$ can be estimated by the inverse of the observed information matrix. The correlation was estimated to be 0.10.

Table 9 gives a summary of the results from the various variance component estimation procedures.

As with the low pressure data, β can be estimated using a form of the jackknife. This should give a variance which is biased toward zero, however, since it treats the δ_i's as fixed effects.

TABLE 9.   SUMMARY OF VARIANCE COMPONENT ESTIMATES

Method	$\hat{\sigma}_\delta^2 \times 10^6$	$\hat{\sigma}_\delta$
Equal weights	60.0	0.00775
Raw weights	277.9	0.01667
Modified weights		
One step	239.3	0.01547
Full iteration	244.0	0.01562
Maximum likelihood	269.0	0.01640

Table 10 is a summary of the different estimates of β, together with estimated standard errors, estimates of β^{-1}, and confidence sets for β^{-1}. The latter were obtained in terms of confidence sets for β, as was done for the low pressure experiment. Bias corrections ranged from 0.1 to 0.7.

SUMMARY

In both experiments the combined analysis, which fit one line to all of the data, produced results which did not agree well with those from the other analyses. Since the values of $C_{xx}^{(i)}$ ranged so widely, use of weighted average produced perceptible drops in the variances, as well as changes in the estimates. In the ouabain experiment, however, the presence of a random effect reduced the advantage of weighting. Comparing the two methods under "raw weights" in the ouabain data, one can see the substantial variance bias which could have resulted from ignoring the

TABLE 10. SUMMARY OF ANALYSES OF OUABAIN EXPERIMENT

Method	b	$\hat{\sigma}$	$\hat{\beta}^{-1}$	95% Confidence Set
Combined	0.03233	0.00364	30.5	(25.3, 39.7)
Simple average				
Fixed effects	0.04688	0.00525	21.1	(17.5, 27.3)
Sample	0.04688	0.00558	21.0	(16.9, 28.9)
Raw weights				
Fixed effects	0.04124	0.00310	24.1	(21.1, 28.4)
Random effects	0.04124	0.00681	23.6	(18.3, 35.9)
Modified weights				
One step	0.04594	0.00537	21.5	(17.7, 28.2)
Full iteration	0.04581	0.00540	21.5	(17.7, 28.4)
Maximum likelihood	0.04591	0.00558	21.5	(17.6, 28.6)
Jackknife	0.04349	0.00779	22.3	(17.0, 35.4)

random nature of the effect. In both experiments the jackknife produced the largest variance estimate.

Since the previous investigator's value of 29 for β^{-1} was obtained in a combined analysis, an irrelevant factor confounded the estimate, and comparisons with the present results are difficult. One can at least note, though, that a value of 29 is in good agreement with the combined analysis from the ouabain experiment, but this is not true of the low pressure experiment.

It is interesting to see that the two experiments are not in good agreement on the values of β^{-1}. This suggests that the method of changing equilibrium might also affect the sodium pumping efficiency.

REFERENCES

Kendall, M. G. and A. Stuart (1961). The Advanced Theory of Statistics, Vol. 2. New York: Hafner.

Miller, R. G. (1974). An unbalanced jackknife. Ann. Statist., 2, 880–891.

Pearson, E. S. and H. O. Hartley (1954). Biometrika Tables for Statisticians, Vol. 1. Cambridge: University Press.

Searle, S. R. (1971). Linear Models. New York: Wiley.

Snedecor, G. W. and W. G. Cochran (1967). Statistical Methods, 6th ed. Ames: Iowa State Press.

Kidney	Sodium Reabsorption[a]	Oxygen Consumption[b]
1	9171	4698
	6309	4900
	4886	4486
	5297	4346
	4604	4165
2	6987	4096
	7154	4437
	5584	3218
	4895	3396
3	4307	5224
	4197	5349
	2403	4426
	1086	4458
	2320	4367
4	9685	5712
	9406	5294

Kidney	Sodium Reabsorption	Oxygen Consumption
	4292	3789
	4448	4736
	4535	5126
5	8160	5191
	9790	5564
	2560	4078
	7350	4290
	8350	5024
6	7888	6558
	7868	5938
	386	5205
	3789	7197
	2147	4318
7	4171	5865
	6509	6878
	1456	4618

185

Kidney	Sodium Reabsorption	Oxygen Consumption	Kidney	Sodium Reabsorption	Oxygen Consumption
	2078	4618		4682	4770
	3475	3699		2769	4361
8	8063	4645		2927	3380
	6460	4099		1803	4225
	1177	3452	10	7483	4139
	1457	3206		6308	4215
	1768	3206		4035	3399
				4230	3399
9	4943	5561		4719	3332

[a]Unit of sodium reabsorption is 10^{-2} µEq/min.

[b]Unit of oxygen consumption is 10^{-3} µmoles/min.

APPENDIX 2: OUABAIN DATA

Kidney	Dry Weight[a]	Sodium Reabsorption[b]	Oxygen Consumption[c]	Kidney	Dry Weight	Sodium Reabsorption	Oxygen Consumption
1	2722	12983	6016			1736	3667
		13150	5481	4	2619	4228	3772
		3710	2727			5236	4646
		5631	3266			2677	2299
		5213	3315			3205	3197
		5517	3840			5113	3260
2	3128	6972	5739			4515	3260
		7684	6109	5	2959	10448	4886
		2624	1388			10063	5830
		2647	3304			4748	3120
		3953	2706			5663	4437
		4234	3065			5416	3908
3	3434	4604	4980			4916	3984
		4404	4214	6	2765	5893	4597
		2070	3232			4910	4291
		3168	3134			2712	2964
		3023	3706				

Kidney	Dry Weight	Sodium Reabsorption	Oxygen Consumption	Kidney	Dry Weight	Sodium Reabsorption	Oxygen Consumption
7	3016	2704	1902	10	2576	4864	4450
		2438	2925			1756	2790
		2827	2758			1070	3370
		5176	4698			8340	4050
		5329	4698			6110	3640
		1571	2438			4450	2700
		2690	3291			2850	3860
		1399	2309	11	3660	6924	4960
		2877	3316			6048	4960
8	3184	3476	4037	12	2705	5230	4880
		3370	4934			3457	4140
		3500	2741			7321	5680
		3160	3795			8231	5850
		2557	3044			6080	3570
		2514	3275			4088	2820
9	2728	5433	5850				

APPENDIX 2 (Continued)

Kidney	Dry Weight	Sodium Reabsorption	Oxygen Consumption	Kidney	Dry Weight	Sodium Reabsorption	Oxygen Consumption
13	2844	4704	4860			4216	3550
		5722	5520			4067	3540
		5761	5490	16	2775	7685	5220
		4134	5590			6834	3340
14	3032	6339	4870			3278	2380
		5767	4960			3264	2260
		3228	2480	17	3272	9002	4280
		3012	2970			8013	4160
15	2887	6338	4930			4723	3260
		8569	5130			2320	2840

[a]Unit of dry weight is 10^{-1} mg.

[b]Unit of sodium reabsorption is 10^{-2} μEq/min.

[c]Unit of oxygen consumption is 10^{-3} μmoles/min.

EVALUATING LABORATORY MEASUREMENT TECHNIQUES

SUE LEURGANS

<u>Medical Problem</u>. Comparison of two methods of measurement in the clinical laboratory.

<u>Medical Investigator</u>. Steven Levine, Stanford University Hospital.

<u>Statistical Procedures</u>. Regression analysis, tolerance intervals, structural relationship, and components of variance.

INTRODUCTION

The statistical problems discussed in this report arose in the Stanford Medical Center Clinical Laboratories. Two kinds of problems are discussed. One problem is the comparison of two methods of measuring the same quantity. The other problem involves the control charts used to monitor any particular method.

Since both of these problems occur frequently, the techniques of analysis are of more interest than the analysis of any particular data set. The first problem is attacked in several stages, each stage relaxing an earlier assumption. We illustrate the methods with calculations for some glucose data.

The first problem arises whenever a new method of making measurements is being developed or whenever a new kind of machine is purchased. Usually there is an accepted method or machine, called the reference method. The other method is called the new method, or test method. Since most hospitals and laboratories generally have some institutional inertia, the new method will be judged by the accuracy with which it reproduces the results of the reference method. The new method will be deemed acceptable if it deviates from the reference method at certain prespecified levels of concentration by less than the corresponding allowable error. This formulation of the problem is presented by Westgard et al. (1974) in the language of clinical chemistry. When carefully stated, this approach searches for a tolerance interval for the new method, given the value of the reference method. We focus on comparing the methods on patients, and ignore analyses of more controlled samples. These analyses include recovery studies (which add a known amount of the substance being measured and observe the proportion recovered) and interference studies (which add substances that may distort the results). While these studies are important in evaluation of the methods, they cannot replace patient comparison studies. Nor are the statistical analyses the same. This work focused on patient comparison studies because the method of analysis recommended in the clinical chemistry literature was found to be defective.

The tolerance interval approach is addressed to the question of whether the test method approximates the reference method accurately enough. There is another, more symmetric view of the problem. Since no method is perfect in practice, the results from both methods are measurements that combine the true values with error. The clinical chemists want to know if the underlying true values implicit in each method are identical. If we were to assume the two values are linearly related, the appropriate analysis would compare the underlying intercept and slope with 0 and 1, respectively.

The second problem arises, among other times, in the collection of data for the examination of the first problem. Clinical measurements often involve complicated machines. Standard preparations are used to check that the method is running correctly. The values of the standards are plotted on control charts. Current practice in the Stanford Clinical Laboratories is to construct these charts every month by plotting a mean level plus and minus 2 standard deviations, where both the mean level and the standard error are calculated from the collection of all measurements of the same standard in the previous month. This is recognized to be ad hoc, and more sophisticated procedures are desired.

THE DATA SETS

Background

The data which are used to illustrate the procedures come from two methods of measuring the concentration of glucose in the blood. The reference method uses a Technicon SMA 6/60. Glucose is one of six substances that this machine can measure. The 6/60 is centered around a colorimeter. The test for glucose uses a substance that changes color in the presence of glucose. Like many color tests, this test is vulnerable to interference. In this case, uric acid neutralizes the effect of glucose on the substance. The test method used the System I. This method uses an enzymatic test, which is more specific to glucose.

Patient Comparison Study

This study was run in January 1976, at the Clinical Laboratories of the Stanford Medical Center. For each of the 46 different patients, the glucose concentration was measured by both of the methods above. The data obtained are given in Appendix 1.

For the duration of this report, X_i is the glucose level determined by the 6/60 for the i-th patient, and Y_i is the patient's System I measurement. Thus X is the reference method, and Y the new method.

Control Chart Data

These numbers are all of the numbers that were recorded on control charts of glucose measurements on these two machines during a four-week period that matches that of the patient comparison study. We shall describe in detail how the measurements are made.

The 6/60 obtains samples from a wheel which can hold 40 different samples. (Batch will be a synonum for wheel.) Each wheel contains the following sequence of samples: 3 samples of calibration serum, 10 patient samples, 2 calibration samples, 10 patient samples, 2 calibration samples, and 10 patient samples. Before the machine starts sampling from the wheel, the tubing is flushed with water. This also sets a base level. Approximately 10 minutes after the samples start through the machine, the results begin to appear. A value is plotted on the appropriate axis. Since the samples are processed at the approximate rate of one sample every 10 seconds, these values are plotted quite rapidly. Thus, the first calibration sample is just observed to

see what portion of the axis is crossed. The operator then
adjusts the machine so that the second value is plotted at
exactly the correct value (which is, of course, known, because
this is a calibration sample). In the first triplet there is a
third calibration sample. This is observed to check that the
adjustment was correct.

There are two levels of control samples. The low level serum
has a glucose concentration near 74 mg/dl and the high level
serum has a concentration near 295 mg/dl. The calibration serum
has a concentration near 190 mg/dl. One sample of each control
serum is included in each wheel. The controls are treated as
though they came from patients. That is, 2 of the 30 patient
samples referred to above are controls. The controls are placed
quasi-randomly among the patients. (The person who sets up the
wheels is not the person who operates the machine.)

For the 6/60, the calibration samples are a secondary clinical
standard. This means that the standard is serum based. The
serum is obtained either from outdated blood in blood banks or
from blood bought by the drug company from individuals. The
serum is pooled in large lots. The company making the calibration serum then determines the amount of glucose in the serum by
sending samples to roughly 10 reference laboratories for assay.
The mean of these values is then supplied to the user with the
calibration serum.

The System I procedure is similar. The primary differences
are that batches are roughly half the size of 6/60 batches (that
is, 15-20 patients/batch) and that the calibration samples are a
primary clinical standard. Primary clinical standards are water
based (rather than serum based). The concentration of glucose is
determined very precisely by weighing dry glucose before the
standard is prepared. Primary clinical standards are certified
by the National Bureau of Standards and are distinctly more reliable than secondary standards. Use of a primary clinical standard also means that the calibration serum for the 6/60 can be
used as a (medium-level) control serum for the System I.

The data available are all of the control values that were
recorded during the daytime block (roughly 7 a.m. to 4 p.m.) during a four-week period in January and February, 1976. This period includes that of the patient comparison study. The values
are recorded opposite the day number. These data are in Appendix
2. It is believed that some control measurements were not
recorded.

TOLERANCE INTERVALS

Westgard et al. (1974) discuss the problems faced by a clinical chemist who wishes to evaluate his new method. They assume
that some medical authorities have stated that measurements at
particular concentrations should be within some stated error.
Thus for glucose there are two critical concentrations: 50 and
120 mg/dl. The stated allowable error is 10 mg/dl at both concentrations (Westgard et al., 1976).   50 and 120 mg/dl are near
the boundaries of glucose levels for normal adults.
     Westgard et al. (1974) suggest that the new method be deemed
acceptable if the 95% upper limits of the error of the new method
at the critical concentrations are less than the corresponding
stated allowable error rates. They also suggest that the method
be rejected if the 5% lower limits exceed the stated allowable
errors, and that no decision be made if neither acceptance or
rejection occurs. However, their calculations are in error. The
lower limits cannot be repaired. Tolerance intervals seem to be
an appropriate way to replace the upper limits.
     This section describes the first simple analysis. Before pre-
senting the tolerance intervals, the standard linear model will
be applied. That is, we shall assume

$$Y_i = \alpha + \beta X_i + E_i, \qquad E_i \sim n(0, \sigma^2), \qquad i = 1,\ldots,N, \quad (1)$$

where σ^2 is unknown and the E_i are independent.

     The least squares estimates of and confidence intervals for
the parameters are given in Table 1. The associated ANOVA table
is included because the estimate of residual error is used subse-
quently.
     Since the line $\alpha = 0$ and $\beta = 1$ plays a distinguished role in
this analysis, it is natural to test the corresponding hypothesis. The test statistic is

$$\frac{1}{\hat{\sigma}^2} (\hat{\alpha}, \hat{\beta} - 1) \, \underset{\sim}{X}'\underset{\sim}{X} \begin{pmatrix} \hat{\alpha} \\ \hat{\beta} - 1 \end{pmatrix} \overset{H_0}{\sim} F_{2,\ N-2} \, . \tag{2}$$

This statistic has the value 435.8, largely because the slope
appears to be less than 1. So we reject the hypothesis.
     A 95% confidence interval for the intercept is

TABLE 1. SUMMARY OF LEAST SQUARES CALCULATIONS

ANOVA Source	DF	SS	MS
Regression	1	310,090.96	
Residual	44	1,384.87	31.474
Total corrected for mean	45	311,475.83	
Mean	1	822,658.17	
Total	46	1,140,134.00	

Fitted regression line: $\hat{Y} = 2.748 + 0.912\ X$.
Estimated standard deviation: $\hat{\sigma} = 5.610$.

$$\hat{\alpha} \pm t_{N-2,\ 0.975} \left[\frac{\Sigma\ X_i^2}{N\ \Sigma(X_i - \overline{X})^2} \right]^{1/2} S \ , \tag{3}$$

and for the slope, the interval is

$$\hat{\beta} \pm t_{N-2,\ 0.975} [\Sigma(X_i - \overline{X})^2]^{-1/2} S \ , \tag{4}$$

where $t_{N-2,\ 0.975}$ is defined by

$$P\{t_{N-2} \leq t_{N-2,\ 0.975}\} = 0.975 \ . \tag{5}$$

Using the glucose data, (3) is 2.748 ± 3.158, or (−0.410, 5.906). The interval (4) reduces to 0.912 ± 0.018 = (0.893, 0.931).

The residuals are discussed in a later section. We shall just note here that there is no evidence of an important quadratic term.

Tolerance Intervals: Definition

We say that we have a tolerance interval if we have a method of constructing intervals such that with probability γ, the interval we construct captures more than P of the probability under the true distribution.

To express this more mathematically, consider the class of tolerance intervals

$$Z \pm kS , \tag{6}$$

where Z is a random variable, k is a constant, and S is the standard deviation. Z will be chosen so that

$$Z \sim \hbar(\mu, \sigma^2/m) , \tag{7}$$

$$nS^2 \sim \sigma^2 \chi_n^2 , \tag{8}$$

independent of Z, where m is known.

$$A = \int_{Z-kS}^{Z+kS} \frac{1}{\sigma\sqrt{2\pi}} e^{-(x-\mu)^2/2\sigma^2} dx \tag{9}$$

is the amount of probability captured by the intervals. A is a function of k, m, n, and S. Since A does not depend on μ or σ, the calculations can be done with $\mu = 0$ and $\sigma = 1$. If we now fix m and n,

$$F(P,k) = P\{A \geq P\} \tag{10}$$

is the quantity we wish to be greater than γ.

Wallis Intervals

Wallis (1951) sketches an argument that introduces $F(P,k|Z)$, expands this in a Taylor series around $Z = 0$, takes expected values term by term, and, comparing this with the original Taylor series, shows that

$$F(P,k) = F(P,k|1/\sqrt{m}) + O(m^{-2}) . \tag{11}$$

This approximation is used to reduce the problem to solving equation (12) for r, where $\phi(t)$ is the unit normal density:

$$\int_{1/\sqrt{m} - r}^{1/\sqrt{m} + r} \phi(t) dt = P . \tag{12}$$

Bowker (1946, 1947) derived a Taylor series for r, and tabled some relevant constants for selected P and γ. Some of these constants include approximations of χ^2 percentage points. The resulting formula for k is

$$k = K(P) \left[1 + \frac{1}{2m} - \frac{2K^2(P) - 3}{24m^2}\right] \left[\frac{n}{\chi^2_{n,\gamma}}\right]^{1/2} , \tag{13}$$

where

$$\int_{-K(P)}^{K(P)} \phi(z) \, dz = P ,$$

$$\tag{14}$$

$$P\{\chi^2_n \leq \chi^2_{n,\gamma}\} = \gamma .$$

In the context of a tolerance interval for Y at $X = x_0$, x_0 known,

$$Z = \hat{\alpha} + \hat{\beta} x_0 , \tag{15}$$

$$\frac{1}{m} = \frac{1}{N} + \frac{(x_0 - \overline{X})^2}{\displaystyle\sum_{i=1}^{N} (X_i - \overline{X})^2} . \tag{16}$$

Table 2 gives the tolerance intervals for selected x_0.

Bonferroni Intervals

Another approach uses Bonferroni's inequality

$$P(A \cap B) \geq 1 - P(A^c) - P(B^c) . \tag{17}$$

To apply this inequality, we need to select the sets A and B appropriately. A will be a set on which our estimate of $E(Y|X)$ is not wildly far from the true value, relative to the sample standard deviation. B is a set which excludes misleadingly small estimates of the standard deviation. Definitions (18) and (19) are more precise:

$$A = \left\{ m^{1/2} \frac{|\hat{\alpha} + \hat{\beta} x_0 - (\alpha + \beta x_0)|}{S} \leq t_{n, \, 1-(1-\gamma)/4} \right\} , \tag{18}$$

TABLE 2.   BONFERRONI AND WALLIS TOLERANCE INTERVALS FOR SELECTED
VALUES OF x_0

x_0	Bonferroni Interval		Wallis Interval		$\hat{\alpha} + \hat{\beta}x_0$
	Lower	Upper	Lower	Upper	
30	21.03	39.21	22.78	37.46	30.12
50	39.45	57.28	41.06	55.67	48.37
70	57.86	75.38	59.34	73.89	66.62
90	76.24	93.50	77.62	92.12	84.67
110	94.59	111.65	95.88	110.35	103.12
130	112.89	129.83	114.14	128.59	121.36
150	131.15	148.07	132.39	146.84	139.61
170	149.36	166.36	150.63	165.09	157.86
190	167.52	184.69	168.87	183.35	176.11
210	185.65	203.06	187.09	210.62	194.36
230	203.75	221.46	205.31	219.90	212.61
250	221.84	239.87	223.53	238.18	230.86

$$B = \left\{ nS^2 > \sigma^2 \chi^2_{n, (1-\gamma)/2} \right\} . \tag{19}$$

$P(A^C) = P(B^C) = (1-\gamma)/2$, which implies

$$P(A \cap B) \geq \gamma . \tag{20}$$

Now we consider expression (21), which gives the distances
between the endpoints of the interval that contains the central P
proportion of the normal distribution and the least squares estimator of the expected value.

$$\left| \alpha + \beta x_0 \pm K(P) \sigma - (\hat{\alpha} + \hat{\beta}x_0) \right| . \tag{21}$$

We separate the $K(P) \cdot \sigma$ term and first apply the triangle inequality and then the definitions of A and B to demonstrate that on
$A \cap B$, Δ is an upper bound for (21).

$$\Delta = S \cdot t_{n,\ 1-(1-\gamma)/4}\ (m)^{-1/2} + K(P)S \left[\frac{n}{\chi^2_{n,\ (1-\gamma)/2}} \right]^{1/2} . \quad (22)$$

Now, by (20)

$$P((21) \leq \Delta) \geq \gamma . \quad (23)$$

But the event in (23) is equivalent to the event

$$\hat{\alpha} + \hat{\beta}x_0 - \Delta \leq \alpha + \beta x_0 \pm K(P)\ \sigma \leq \hat{\alpha} + \hat{\beta}x_0 + \Delta . \quad (24)$$

Thus on A ∩ B, the interval

$$\hat{\alpha} + \hat{\beta}x_0 \pm \Delta \quad (25)$$

contains more than P of the probability. Since $P(A \cap B) \geq \gamma$ the interval (25) is a tolerance interval of the desired kind. See Table 2 for selected values. These intervals are uniformly wider than the Wallis intervals, as can be seen in Figure 1.

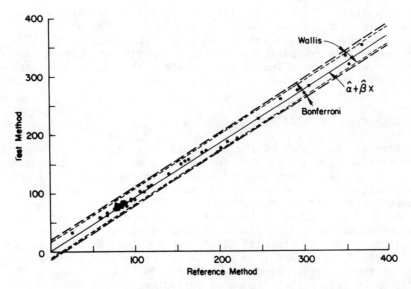

Figure 1.   Nonsimultaneous tolerance intervals

Comments

There are several differences between the procedures. Most obvious is the fact that the Bonferroni interval is a conservative procedure, while the Wallis interval is an approximation to an exact procedure. Also, the Bonferroni procedures enable one to use two distinct estimators of σ. The same formal manipulations are possible if the estimators of σ in (18) and in (19) are different. The expression for Δ will involve both estimators, each multiplied by a function of an appropriate percentage point. However, while two such estimators may well be available, one or the other or some pooled estimator will probably be preferred.

A more subtle difference is that the Bonferroni interval is constructed to capture the P proportion of the true distribution symmetrically (i.e., P/2 of the part of the true probability below the mean and P/2 of the part of the true distribution above the mean), while the Wallis interval is just constructed to capture any connected P portion of the probability.

All of these tolerance intervals are nonsimultaneous. That is, the distribution theory is valid only when these tolerance intervals are constructed for one value of x_0. Thus the theory is most applicable when there is only one critical concentration level.

Tolerance intervals for all values of X that are simultaneously valid can also be constructed. Several methods are described in Lieberman and Miller (1963). The Bonferroni approach gives an interval of the same form as (25). The percentage points are slightly different, causing the simultaneous Bonferroni tolerance interval to be wider than the nonsimultaneous Bonferroni tolerance interval at every value of X.

EXAMINING THE VARIANCE ASSUMPTION

Figure 2 is a plot of the residuals from the linear fit of the previous section. This plot does seem to contradict the assumption of constant error variance. Fortunately, the control chart information can be used to obtain more information about the variance, which will be independent of the linear effect fitted.

Table 3 is a table of the various estimates of the standard deviation at particular levels of concentration. Six of the estimates are from Stanford Clinical Laboratory data, three for each method. Five are from a table in Westgard et al. (1976). One of these was obtained from a "blind duplicates" study. In this study two samples from each patient were analyzed separately. All of the other estimates are from control sera.

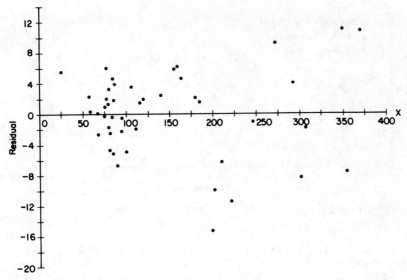

Figure 2.   Residuals from linear regression of System I on 6/60.
Line: Y = 2.748 + (0.912)X.

The estimated standard deviations are plotted against the sample
means in Figure 3.

TABLE 3.   MEASUREMENT ERRORS

Measured at Stanford				Measured by Westgard et al. (1976)	
6/60		System I			
Mean	Standard Deviation	Mean	Standard Deviation	Mean	Deviation
74	2.6	74	2.3	87	2.23
208	3.1	189	3.6	83	1.62
295	6.1	270	5.7	212	4.72
				354	5.59

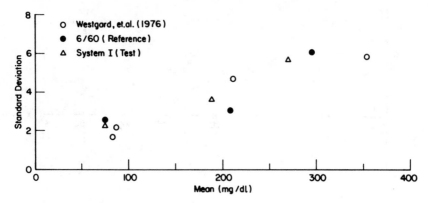

Figure 3.   Standard deviations versus means

These data also contradict the assumption that the underlying variance is constant. The variance increases with X. It seems more plausible that the standard deviation of Y at X = x is some constant multiple of x.

There are at least two models that would imply this last assumption. They can be presented symbolically as

$$Y_i = \alpha + \beta X_i + E_i, \qquad E_i \sim \hbar(0, \sigma^2 X_i^2) ; \qquad (26)$$

$$\ln Y_i = \alpha' + \beta' \ln X_i + E_i', \qquad E_i' \sim \hbar(0, \sigma^2) . \qquad (27)$$

The model in (27) is equivalent to

$$Y_i = e^{\alpha'} X_i^{\beta'} e^{E_i'} . \qquad (28)$$

Thus (26) is a model with additive errors, while (27) is a model with multiplicative errors.

The Logarithmic Analysis

The logarithmic analysis is a straightforward application of earlier methods. Table 4 presents some statistics from the regression of V on U, where V = ln Y and U = ln X. Figure 4 is the corresponding residual plot. It is immediately apparent that

TABLE 4.   RESULTS FROM REGRESSION OF V ON U

Source	DF	SS	MS
Regression	1	14.4102	
Error	44	0.1158	2.632×10^{-3}
Total Corrected for Mean	45	14.52605	
Mean	1	1031.13684	
Total	46	1045.66289	

Fitted regression line: $\hat{V} = 0.177 + 0.951\ U$.

Estimated standard deviation: $\hat{\sigma} = 5.13 \times 10^{-2}$.

the point with the smallest value of U and the smallest value of V has by far the largest residual. Therefore the analysis was repeated, omitting this point. The results are summarized in Table 5, and the residual plot in Figure 5.

We now repeat the tolerance interval calculations of the previous section. We continue to delete the one point that appeared to be an outlier. The resulting tolerance intervals are then transformed back to the original scale. The resulting Bonferroni and Wallis intervals are superimposed on the data and the estimated line in Figure 6. Table 6 gives selected values from these curves.

Weighted Least Squares

The computations for the weighted least squares line are summarized in Table 7. The weighted residuals, which have equal variance under the model, are Figure 7. The formula for the weighted least squares Bonferroni tolerance interval at x_0 is

$$\hat{\alpha} + \hat{\beta}x_0 \pm \Delta(x_0) \ , \tag{29}$$

where

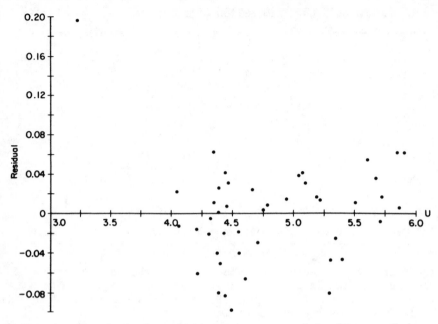

Figure 4.   Residuals from linear regression of ln (System I) on
ln (6/60).   Line: V = 0.177 + (0.951)U.

$$\Delta(x_0) = S \left[\left(\frac{1}{W} + \frac{(\tilde{X} - X_0)^2}{\Sigma \, w_i (X_i - \tilde{X})^2} \right)^{1/2} t_{n, \, 1-(1-\gamma)/4} \right.$$

$$\left. + K(P) \left(\frac{n}{\chi^2_{n, \, (1-\gamma)/2}} \right)^{1/2} \right],$$

(30)

$$W = \Sigma \, w_i \, ,$$

$$\tilde{X} = \Sigma \, X_i w_i / W \, .$$

Selected values of (29) are listed in Table 8.

TABLE 5. RESULTS FROM REGRESSION OF V ON U, ONE OUTLIER DELETED

Source	DF	SS	MS
Regression	1	12.728	
Error	43	0.069	1.599×10^{-3}
Total Corrected for Mean	44	12.797	
Mean	1	1021.074	
Total	45	1033.871	

Fitted regression line: $\hat{V} = 0.059 + 0.974\ U$.

Estimated standard deviation: $\hat{\sigma} = 3.999 \times 10^{-2}$.

Conclusions

The logarithmic analysis seems to be the most satisfactory. However, even for this analysis and for the weighted least squares, the tolerance intervals are not as tight around 50 and 120 as the medical standards require. Thus on the basis of this data, the new glucose method does not quite pass the stringent requirements of the tolerance interval version of Westgard's formulation.

STRUCTURAL RELATIONSHIP

The model we discuss here is called the structural model by Kendall and Stuart (1961). We follow them in denoting the true, unobserved values by X and Y. The model asserts that

$$Y = \alpha + \beta X . \tag{31}$$

We observe ξ and η, where

$$\xi = X + \delta ,$$

$$\eta = Y + \epsilon . \tag{32}$$

δ and ϵ are errors with means zero and variances σ_δ^2 and σ_ϵ^2, respectively. We assume that all the errors are independent.

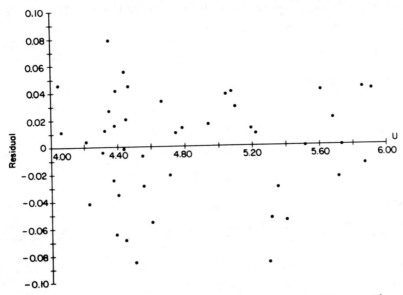

Figure 5.   Residuals from linear regression of ln (System I) on ln (6/60), one point deleted.   Line: V = 0.059 + (0.974)U.

Thus our data will be N pairs of (ξ_i, η_i). The model also postulates that the true X's of our sample are themselves a sample from a population with mean μ and variance σ_X^2.

This model seems natural in this situation. Each person has a true underlying reference method value, which is one value from all those which humans can assume. The person's true test method value is assumed to be a linear function of his reference method value. Both of these true values are then observed with error. This error is now assumed to have variance independent of the underlying true values.

Many estimators have been derived for this situation. We consider two groups: maximum likelihood estimators and instrumental variable estimators. In both cases it is assumed that the ξ_i's and the η_i's are measured from their observed means. This makes the intercept estimate zero. In the original coordinates, if $\tilde{\alpha}$ estimates α and $\tilde{\beta}$ estimates β, we will find $\tilde{\alpha}$ from

$$\tilde{\alpha} = \overline{\eta} - \tilde{\beta}\ \overline{\xi}\ .$$

(33)

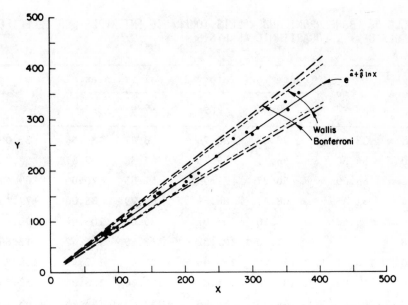

Figure 6.   Nonsimultaneous tolerance intervals from logarithmic analysis, one point deleted.

The maximum likelihood estimators are for the case in which all distributions are normal. There is a slightly embarrassing consequence of this assumption. This model imposes so much structure that the model is not identifiable. There are six parameters, but the sufficient statistic is a five-dimensional vector. The most natural way to evade this problem is to assume some relationships among the parameters. A common assumption is that λ, the ratio of σ_ϵ^2 to σ_δ^2, is known. The MLE under this assumption is a root of a quadratic.

$$\hat{\beta} = \frac{s_\eta^2 - \lambda s_\xi^2 + [(s_\eta^2 - \lambda s_\xi^2)^2 + 4\lambda s_{\xi\eta}^2]^{1/2}}{2s_{\xi\eta}} \, , \tag{34}$$

where

TABLE 6.   BONFERRONI AND WALLIS TOLERANCE INTERVALS FOR SELECTED
VALUES OF x_0, LOGARITHMIC ANALYSIS

U_0	$x_0 = e^{u_0}$	Bonferroni Interval		Wallis Interval		$e^{\hat{\alpha} + \hat{\beta} \cdot u_0}$
		Lower	Upper	Lower	Upper	
3.5	33.12	28.02	36.76	28.97	35.56	32.09
4.0	54.60	46.12	59.16	47.32	57.67	52.24
4.2	66.69	56.27	71.60	57.55	70.00	63.47
4.4	81.45	68.62	86.69	69.99	85.00	77.13
4.6	99.48	83.60	105.07	85.08	103.24	93.72
4.8	121.51	101.70	127.52	103.40	125.42	113.88
5.0	148.41	123.49	155.06	125.63	152.42	138.38
5.1	164.02	136.00	171.08	138.46	168.04	152.54
5.2	181.27	149.73	188.83	152.60	185.28	168.15
5.3	200.34	164.79	208.47	168.16	204.30	185.35
5.4	221.41	181.34	230.21	185.30	225.28	204.31
5.5	244.69	199.51	254.25	204.18	248.44	225.23

$$S_\eta^2 = \sum_{i=1}^{N} \frac{(\eta_i - \bar{\eta})^2}{N} \, ,$$

$$S_\xi^2 = \sum_{i=1}^{N} \frac{(\xi_i - \bar{\xi})^2}{N} \, , \tag{35}$$

$$S_{\xi\eta} = \sum_{i=1}^{N} \frac{(\xi_i - \bar{\xi})(\eta_i - \bar{\eta})}{N} \, .$$

In view of the previous section, we apply this estimator to the
logarithms of the patient comparison measurements, assuming
$\lambda = 1$.   This assumption seems consistent with Figure 3.   The
results are in Table 9.

TABLE 7.   SUMMARY OF WEIGHTED LEAST SQUARES CALCULATIONS

$$n = 46$$

$$\sum_{i=1}^{n} w_i = 6.12 \times 10^{-3} \qquad \sum_{i=1}^{n} X_i^2 w_i = 12.885$$

$$\sum_{i=1}^{n} X_i w_i = 0.4502 \qquad \sum_{i=1}^{n} Y_i^2 w_i = 9.971$$

$$\sum_{i=1}^{n} Y_i w_i = 0.43675 \qquad \sum_{i=1}^{n} X_i Y_i w_i = 11.292$$

Fitted regression line: $\hat{Y} = 6.903 + 0.876X$

Estimated standard deviation: $\hat{\sigma} = 4.319 \times 10^{-2}$

When normality is assumed and λ is assumed known, the maximum likelihood estimator of σ_δ^2 is

$$\hat{\sigma}_\delta^2 = s_\xi^2 - \frac{S_{\xi\eta}}{\hat{\beta}} . \tag{36}$$

$\hat{\sigma}_\delta^2$ for $\lambda = 1$ is included in Table 9.

The instrumental variable approach does not require normality assumptions. The estimators use an additional random variable ζ, which is correlated with X, but uncorrelated with δ and with $\mathcal{E}$. The estimator of the slope is of the form

$$b = \frac{\sum\limits_{i=1}^{N} \zeta_i \, \eta_i}{\sum\limits_{i=1}^{N} \zeta_i \, \xi_i} . \tag{37}$$

Kendall and Stuart (1961, Vol. II, p. 398) show that b is a consistent estimator for β, unless X and ζ are asymptotically (in N) uncorrelated.

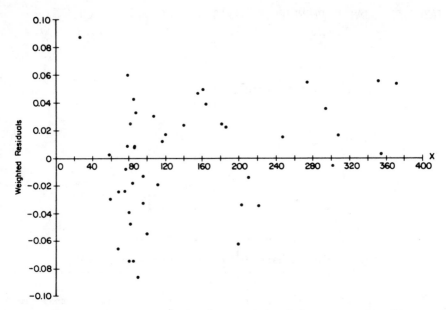

Figure 7.   Weighted residuals from weighted least squares line.

We shall apply two instrumental variable estimators.   Each
instrumental variable is a difference of two indicator functions.
For the first estimator, one indicator variable is one only if
the observed ξ_i is above the sample median, and the other indicator variable is one if the observed ξ_i is below the sample

median.   This estimator, originally suggested by Wald (1940), was
one of the first instrumental variables estimators suggested.
The indicator functions for the second estimator, due to Bartlett
(1949), indicate the region containing the largest third of the
ξ_i's and the region containing the smallest third of the ξ_i's.

The glucose data calculations are summarized in Table 10.   These
calculations use the logarithms of the concentrations.
     The structural analysis approach seems to be the most natural
of the approaches, probably because the assumptions seem most
plausible.   While this analysis does not fit into the Westgard
approach, it does seem to be an informative way to summarize the
data.

TABLE 8.   BONFERRONI TOLERANCE INTERVALS FOR SELECTED VALUES OF x_0, WEIGHTED LEAST SQUARES

x_0	Bonferroni Intervals		$\hat{y}$
	Lower	Upper	
50	44.22	57.22	50.72
70	59.84	68.25	76.65
90	75.25	96.30	85.77
110	90.46	116.14	103.30
130	107.59	134.84	121.21
150	123.42	155.25	139.33
170	139.18	175.72	157.45
190	154.92	196.22	175.57
210	170.64	216.74	193.69
230	186.35	237.27	211.81
250	202.05	257.80	229.93

CONTROL CHART: COMPONENTS OF VARIANCE

The clinical chemists wish to know whether their control charts are using the appropriate standard deviation. The standard deviation they have been using is obtained from the sum of

TABLE 9.   MAXIMUM LIKELIHOOD ESTIMATES, $\lambda = 1$

$$\bar{\xi} = 4.829 \qquad \hat{\beta} = 0.9767$$
$$\bar{\eta} = 4.763 \qquad \hat{\alpha} = 0.0469$$
$$s_{\xi}^2 = 0.298$$
$$s_{\eta}^2 = 0.284 \qquad \hat{\sigma}_{\delta}^2 = 1.08 \times 10^{-3}$$
$$s_{\xi\eta} = 0.290$$

TABLE 10. INSTRUMENTAL VARIABLE ESTIMATORS

Two-Group Instrumental Variable		Three-Group Instrumental Variable	
$b_1 = \dfrac{\bar{\eta}_2 - \bar{\eta}_1}{\bar{\xi}_2 - \bar{\xi}_1}$	$b_1 = 0.9766$	$b_2 = \dfrac{\bar{\eta}_3 - \bar{\eta}_1}{\bar{\xi}_3 - \bar{\xi}_1}$	$b_2 = 0.9708$
	$a_1 = 0.0474$		$a_2 = 0.0754$
$\bar{\xi}_1 = 4.365$	$\bar{\eta}_1 = 4.311$	$\bar{\xi}_1 = 4.308$	$\bar{\eta}_1 = 4.255$
$\bar{\xi}_2 = 5.304$	$\bar{\eta}_2 = 5.228$	$\bar{\xi}_2 = 4.664$	$\bar{\eta}_2 = 4.609$
		$\bar{\xi}_3 = 5.515$	$\bar{\eta}_3 = 5.426$

squares computed by listing all the measurements made by a particular machine on a specific serum, subtracting the grand mean,
and summing the squares of these numbers. To see whether day
information might be useful, a components of variance approach
was tried.

In this approach, Z_{ij}, the j-th observation on the i-th day,
is modeled as

$$Z_{ij} = \mu + A_i + E_{ij}, \qquad \begin{array}{l} i = 1,\ldots,a, \\ \\ j = 1,\ldots,n_i, \end{array} \qquad \Sigma n_i = N , \qquad (38)$$

where the A_i are independent with mean zero and unknown variance
σ_a^2 and the E_{ij} are independent, with mean zero and unknown variance σ_e^2. The E_{ij} and A_i are assumed to be mutually independent.

Unfortunately, the data available are unbalanced. While
several estimators are available, their relative merits are not
clearly understood. We have selected the analysis of variance
estimator for its simplicity. The formulas involved can be
found in Searle (1971, p. 474). These estimators rely on the
first two moments, so normality is not crucial. However, calculations of the variances of the estimators require higher
moments. Thus Searle gives formulas for these variances in the
case of normality.

Since the data are sparse as well as unbalanced, there are
many days on which only one value was recorded. It was feared
that the estimators might be sensitive to these isolated values.
If these days are omitted, the estimator of σ_a changes, while
that of σ_e remains the same. Both sets of results are in Table
11. These data demonstrate that these estimators can yield negative estimates of variances.

This analysis does not seem to reveal any definite components
of variance. For some control measurements, $\hat{\sigma}_a^2$ and $\hat{\sigma}_e^2$ are of the
same order. For other measurements, $\hat{\sigma}_a^2$ is small compared to $\hat{\sigma}_e^2$.
In one case, the comparison changes when isolated measurements
are deleted. It appears that either the data are too sparse for
such methods or that the model is otherwise inappropriate.

As soon as such ideas are entertained, one notices that the
control chart data do have some strange features. For example,
almost half of the values of the low level measurements on the
reference machine are precisely 75. On four days, both of the
two values recorded are 75. These data do not seem particularly

TABLE 11. SUMMARY OF VARIANCE COMPONENTS CALCULATIONS[a]

Level	All Data	Isolated Observation Deleted	All Data	Isolated Observation Deleted
Low				
$\hat{\sigma}_e^2$	6.85 2.84	6.85 2.84	2.33 0.68	2.33 0.68
$\hat{\sigma}_a^2$	0.14 1.46	-0.0457 1.12	3.55 2.5	3.46 3.6
N a	61 28	54 21	41 25	29 13
Medium				
$\hat{\sigma}_e^2$			18.347 56	18.347 56
$\hat{\sigma}_a^2$	This serum is the clinical standard for this machine		-3.19 33	-3.63 18
N a			37 25	23 11
High				
$\hat{\sigma}_e^2$	21.646 39	21.646 39	21.65 93	21.65 93
$\hat{\sigma}_a^2$	17.799 81	21.976 124	17.946 143	2.867 65
N a	51 27	42 18	36 26	20 10

[a]Below each estimate of a component of variance is the value of the normal theory variance of the estimator of the component of the variance.

suited to a model in which the individual measurements have constant variance about a random day effect. Thus it is plausible
that the estimates may not aid our understanding because the

model may not fit the underlying phenomenon. Or perhaps the only problem is that the number of measurements per day is too small for the large-sample consistency properties of moments estimators to influence the behavior of the estimates.

REFERENCES

Bartlett, M. S. (1949). Fitting a straight line when both variables are subject to error. Biometrics, 5, 207-212.

Bowker, A. H. (1946). Computation of factors for tolerance limits on a normal distribution when the sample is large. Ann. Math. Statist., 17, 238-240.

Bowker, A. H. (1947). Tolerance limits for normal distributions. In Eisenhart, et al., Selected Techniques of Statistical Analysis. New York: McGraw-Hill, 95-110.

Kendall, M. G. and A. Stuart (1961). The Advanced Theory of Statistics. London: Griffin, Vol. II, Chap. 29, 357-418.

Lieberman, G. J. and R. G. Miller, Jr. (1963). Simultaneous tolerance intervals in regression. Biometrika, 50, 155-168.

Owen, D. B. (1962). Handbook of Statistical Tables. Reading, Mass.: Wesley.

Searle, S. R. (1971). Linear Models. New York: Wiley.

Wald, A. (1940). The fitting of straight lines if both variables are subject to error. Ann. Math. Statist., 11, 284-300.

Wallis, W. A. (1951). Tolerance intervals for linear regression. Proceedings of the Second Berkeley Symposium on Mathematical Statistics and Probability. Berkeley: University of California Press, 43-51.

Westgard, J. O., R. N. Carey, and S. Wold (1974). Criteria for judging precision and accuracy in method development and evaluation. Clin. Chem., 20, 825-833.

Westgard, J. O., R. N. Carey, D. H. Feldbruegge, and L. M. Jenkins (1976). Performance studies on the technicon "SMAC" analyzer: precision and comparison of values with methods in routine laboratory service. Clin. Chem., 22, 489-496.

APPENDIX 1: PATIENT COMPARISON DATA

Number i	Glucose Measurement Reference Method X_i	Test Method Y_i	Rank of X_i	Rank of Y_i
1	155	150	30	30
2	160	155	31	31
3	180	169	33	34
4	80	79	12	16
5	106	103	25	26
6	185	173	34	35
7	139	132	29	29
8	273	261	40	40
9	220	192	38	38
10	293	274	41	42
11	350	333	44	45
12	79	73	10	9
13	67	64	4	5
14	202	177	36	36
15	57	57	2	3
16	354	318	45	44
17	300	268	42	42
18	370	351	46	46
19	198	168	35	33
20	307	281	43	43
21	77	75	8.5	11.5
22	85	75	17	11.5
23	58	56	3	2
24	100	89	24	24
25	80	71	12	7

| Number | Glucose Measurement | | Rank of | Rank of |
i	Reference Method X_i	Test Method Y_i	X_i	Y_i
26	90	78	21	14
27	74	70	6	6
28	95	87	23	22
29	81	74	14	10
30	84	79	15	16
31	25	31	1	1
32	68	62	5	4
33	80	77	12	13
34	75	72	7	8
35	85	85	17	20
36	77	79	8.5	16
37	120	114	28	28
38	115	109	27	27
39	164	157	32	32
40	85	82	17	18
41	86	83	19	19
42	87	86	20	21
43	247	227	39	39
44	94	88	22	23
45	111	102	26	25
46	210	188	37	37

217

APPENDIX 2: CONTROL CHART DATA, REFERENCE METHOD

Day Number	Low Level Raw Values	High Level Raw Values	Day Number	Low Level Raw Values	High Level Raw Values
1	67 72 70	287 290 285	15	80	
2	75 70	300 305	16	74 75	295 286
3	75 70 72	295 290 300	17	74 75 79	290
4	74 76 75 76	297 299 294	18	75 75	300 295 300 295
5	75	295	19	81 72 70 69	305 300
6	72	296	20	77 75 71	296 285
7	75 77	285 285	21	70	295
8	77 77 74	298 298 300	22	75 75	285 285
9	72 78	304 296	23	75 75	280 300
10	75 70	295	24	75 75	285 290
11	76 74 75	303	25	75 71 75	290 285
12	75	292	26	75	295 292
13	75	296	27	75 71 75	292 285
14	75 75	295 298	28	76 75	287

218

APPENDIX 2: CONTROL CHART DATA, TEST METHOD

Day Number	Low Level	Medium Level	High Level
-1		191	278
0		183	
1	70	189	
2		193 188	271 278
3	73	191	276
4	71	187	271
5	69 71	180	258
6	73	194	276
7	74 71 74	185	273
8	72	191	281
9	73 72	189 193	262 270
10	74	189 192	274
11	70 72	189 191	280
12	68		269
13	71		269
14			
15	73 72	183 191	263 269
16	70	186 190	276 276
17	70 70	185 195	273 272
18	71 72 68	189 195 181	275 266
19	72	187	278
20	78	187	280
21			
22	75 72	196 195	271 278
23	72 71	192 190	276 266
24	72 72	189 187	278
25	73 73		268 270
26	75 71	190	279 272
27	68 66 65	186	258
28	70	186	260

219

ESTIMATION PROBLEMS IN MICROCOLONY AUTORADIOGRAPHY

RUPERT G. MILLER, JR.

Problem. Estimate the percentage of proliferating cells in a tumor after irradiation.

Investigator. Robert Kallman, Stanford University.

Statistical Procedures. Binomial distribution; negative binomial distribution; maximum likelihood; method of moments.

BACKGROUND

A current concept in tumor growth kinetics is that a tumor cell is either actively cycling or it is not cycling at all. Cells of the former type are referred to as proliferating (P) cells, and the latter as nonproliferating, quiscent (Q) cells. P cells are more easily damaged by radiation because they pass through the more radiation sensitive S (DNA synthesis) and M (mitosis) phases of their life cycles. An important question in tumor management is whether Q cells are recruited to become P cells when a number of the P cells have been killed by radiation.

A diagram of a tumor cell life cycle is presented in Figure 1. A P cell passes through the phases G_1, S, G_2, and M in sequence. G_1 and G_2 are resting phases. During the S phase the cell is duplicating its DNA material preparatory to division. During mitosis (M) the chromosomes line up, and the cell divides into two daughter cells. Both daughter cells may be P cells, one may be P and the other Q, or possibly both could be Q cells. A daughter P cell repeats the life cycle, but a Q cell enters an

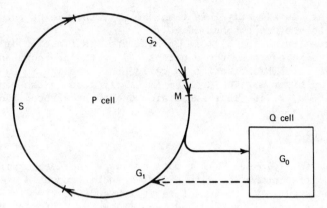

Figure 1.   Diagram for tumor cell cycle.

indefinitely long resting phase G_0. The question under investigation in this series of experiments is whether a Q cell can be stimulated to re-enter the life cycle.

In the past much has been learned about cell kinetics by the percent labeled mitoses (PLM) method. With this technique cells are exposed to a radioactive label, and after the labeling period cell growth is arrested. The cells are covered with a photographic emulsion and kept in a dark room for exposure. Cells in their S phases during the labeling period will have incorporated label and will exhibit grains in the emulsion over their nuclei from radioactive emissions. The fraction of cells in the mitotic stage which exhibit grains is the percent labeled mitoses.

Application of the PLM method to irradiated cell populations can be questioned because it is not possible to morphologically distinguish a surviving clonogenic cell from a doomed cell incapable of division. Thus estimation of the percentage of proliferating cells by the PLM method may be unsound.

A way out of this difficulty is to allow the cells to grow, or not grow, into colonies before arresting growth. Unfortunately, with large colonies containing hundreds of cells (i.e., macrocolonies) the label has become so dilute it is impossible to detect which are labeled colonies. Increasing the amount of label is not a remedy because the radioactivity can reach levels lethal to the cell. The solution adopted for these experiments is to work with small colonies (i.e., microcolonies). In most cases the cells are grown for three days before fixing. Since the average cell cycle for the cells being used in these experiments is slightly less than one day, this means the colonies contain at most about 20 cells.

The statistical problem is to estimate the fraction of colonies which have grown from a parent cell with incorporated radioactive label. This fraction is called the labeling index and is denoted by LI or L. The problem is complicated by the random appearance of grains in the emulsion due to background radiation.

EXPERIMENTS

Mice with implanted EMT6 tumors are used in these experiments. At a specified number of days prior to sacrifice the mouse tumors are irradiated. The amount of radiation and the time interval before sacrifice are varied and constitute the different experimental conditions. Then, for 24 hours before sacrifice the mice are given continuous intravenous infusion of tritiated thymidine label.

After sacrifice some tumor tissue sections are obtained, and tumor cell suspensions are prepared by a trypsinization method. The suspended cells are planted in culture vessels on slides and allowed to grow for varying periods of time. The cells grown for 3 days produce microcolonies and are the primary interest here. However, single cells after only 6 hours, minicolonies after 4 days, and macrocolonies after 13 days are also obtained in some of the experiments.

After the cells are fixed and stained, the slides are dipped in NTB-2 emulsion and stored in the dark for exposure from 23 to 30 days. The number of grains over each cell nucleus in a colony is then counted by eye through a microscope and recorded.

STATISTICAL METHODS FOR SINGLE CELLS

In the measurement of radioactivity, bias in the estimated amount due to background activity can be a worry. In autoradiography background grains can appear in the emulsion even when the cells have not been given a radioactive label. These background grains can be produced by a great many sources as, for example, cosmic rays, nearby radiation, and contamination of the emulsion dip from earlier labeled cells.

For heavy labeling the background can be ignored because it is abundantly clear which cells have incorporated label. However, for nonheavy labeling there is a need to correct for the background in computing the labeling index. For single cells this is an old problem, and there are two standard methods for handling background, the cutoff point method and the Stillström correction.

For the cutoff point approach a number C is selected and only those cells with greater than or equal to C grains are considered labeled. Typically C is 3, 4, or 5. Then,

$$\hat{L} = \frac{\text{number of cells with} \geq C \text{ grains}}{\text{total number of cells}} . \qquad (1)$$

Although this method is often used, it has an obvious difficulty. The cutoff point C should be selected so that there is little probability of a cell with no label having C or more grains. Customarily, a background experiment on cells not exposed to label is run to aid in the selection of C. However, the larger C is made to avoid counting cells with purely background grains, the greater is the chance of not counting cells with a few grains from labeling. Even with scoring large numbers of cells $\hat{L}$ will not converge to the true percentage of labeled cells.

The Stillström (1963) correction stems from a simple probability calculation. Let B, L, and P respectively denote the probabilities that a cell has one or more grains from background alone, from label alone, and from background and/or label. Since the only way a cell can exhibit zero grains is to have none from both background and label, and since the occurrences of background and label grains are probabilistically independent, the probabilities of zero grains must satisfy the equation

$$(1-P) = (1-L)(1-B) . \qquad (2)$$

If binomial estimates $\hat{B}$ and $\hat{P}$ of B and P are substituted from the background only and the label-exposed cell data, the Stillström estimate of the labeling index L obtained by solving equation (2) is

$$\hat{L} = 1 - \frac{1-\hat{P}}{1-\hat{B}} . \qquad (3)$$

The main criticism of the Stillström estimate is that it does not utilize any information in the grain count distribution other than whether a cell has 0 grains or not. However, its simplicity is a great asset, and the extra information is impossible to use unless some assumption about the form of the grain count distribution is warranted.

For weak labeling $\hat{B}$ will be close in size to $\hat{P}$, and this makes the Stillström estimate unstable. If it is safe to assume that

the background and the label grain count distributions are
Poisson, then England, Rogers, and Miller (1973) provide an estimate of L that utilizes more information in the data. However,
the method does require computation of the average number of
grains per cell in the label-exposed and unexposed cultures as
well as the proportions of cells with zero grains in both
cultures.

STATISTICAL ANALYSIS

Barendsen et al. (1973) have run microcolony experiments similar to Kallman's experiments. Their approach to background correction is to use the cutoff point method. The numbers of grains
in the cells of a colony are summed to give the number of grains
for the colony. Colonies consisting of n cells are called
labeled if they contain C_n or more grains. The cutoff points C_n
are derived from the assumption of a Poisson distribution of
background grain counts within a colony of size n. Since the
Poisson distribution will change as the colony size changes, the
cutoff points are allowed to vary with colony size.

David Schoenfeld was the first statistician to examine the
Kallman data. He came to the conclusion that the assumption of a
Poisson distribution for background grains is not justified in
these experiments. Experiment PQ 24 was selected to illustrate
the discussion in this paper, and the results from this experiment on background grains per cell are displayed in Table 1.

Schoenfeld discovered that the negative binomial distribution

$$\binom{i+r-1}{r-1} p^r(1-p)^i , \qquad i = 0,1,2,\ldots \qquad (4)$$

gives a much better fit to the background. This is evident for
PQ 24 from Table 1 where the negative binomial parameters were
estimated by the method of moments $[\hat{p} = \bar{x}/s^2, \hat{r} = \bar{x}^2/(s^2 - \bar{x})]$.
Schoenfeld gave a rational for why the negative binomial distri-
bution might produce a better fit. The number of background
grains per unit area might be Poisson, but the cells can be of
different sizes. If the cell sizes are distributed according to
a gamma distribution, then the number of background grains per
cell will have a negative binomial distribution.

Schoenfeld (1975) felt that the Stillström estimate is too
unstable for colonies so he devised a method for estimating
labeling indices of microcolonies that incorporates the

TABLE 1. NUMBERS OF CELLS WITH DIFFERENT NUMBERS OF GRAINS PER CELL IN GROUP H (NO LABEL) OF EXPERIMENT PQ 24 AND THEIR EXPECTED VALUES (ROUNDED TO THE NEAREST INTEGER) FOR THE POISSON AND NEGATIVE BINOMIAL MODELS

	Grains/Cell				
	0	1	2	3	4
Observed	1513	284	54	10	1
Expected (Poisson $\hat{\lambda} = 0.229$)	1481	339	39	3	0
Expected (negative binomial $\hat{p} = 0.837$, $\hat{r} = 1.177$)	1511	289	51	9	2

assumption of a negative binomial distribution for background counts.

Consider colonies of a fixed size n. A separate analysis is run for each value of n so the dependence on n is notationally suppressed in the following probabilities. Let

$$B_i = P\{i \text{ background grains in a colony}\} \,,$$

$$L_i = P\{i \text{ grains from labeling in a colony}\} \,,$$

and

$$P_i = P\{i \text{ grains in a colony}\}$$

for $i = 0,1,2,\dots$. Then, since the background and label operate independently, the above probabilities convolve in the equations

$$P_i = \sum_{j=0}^{i} L_{i-j} B_j \tag{5}$$

for $i = 0,1,2,\dots$. The labeling index is given by $L = \sum_1^{\infty} L_i = 1 - L_0$.

If the background distribution for grains in a single cell is negative binomial (4) with parameters p and r, then the

background distribution for colonies of size n is negative binomial with parameters p and nr; that is,

$$B_i = \binom{i+nr-1}{nr-1} p^{nr}(1-p)^i .$$

(6)

The parameters p and r are estimated from a separate experiment
(Group H in PQ 24) in which no label is given. The P_i can be
estimated from experimental data (Groups G, A, and B in PQ 24),
but even with estimates of the B_i and P_i there are still too many
unknowns $L_0, L_1, \ldots$ Further restrictions need to be imposed on
the probability structure. Without additional conditions the
maximum likelihood estimate of L is simply the Stillström estimate $\hat{L} = 1 - [(1-\hat{P}_0)/(1-\hat{B}_0)]$.

Schoenfeld added two assumptions that he felt were justified
for the data in this series of experiments. The first is that
the B_i should be negligible for $i > I$ where I is chosen to be 19.
Thus he converts to conditional B_i, given $0 \le i \le I$, and takes
$B_i \equiv 0$ for $i > I$. The second assumption is that, with the exception of L_0, the lower tail of the labeling distribution is essentially flat, that is, $L_1 = L_2 = \cdots = L_I$. Although this assumption may appear severe at first glance, it is not too unreasona-
ble if the central mass of the labeling distribution substantially exceeds I.

Under these assumptions the equations (5) for $0 \le i \le I$ reduce
to

$$P_i = L_0 B_i + L_1 \sum_{0}^{i-1} B_j .$$

(7)

For any labeling experiment the data can be grouped into

N_i = number of colonies with i grains, $0 \le i \le I$ and

N_{I+} = number of colonies with I + 1 or more grains.

(8)

Then, the vector $(N_0, \ldots, N_I, N_{I+})$ of grain count frequencies has
a multinomial distribution with probabilities

$(P_0, \ldots, P_I, 1 - \Sigma_0^I P_i)$ where these probabilities are related to L_0, L_1, and $B_0, \ldots, B_I$ through (7).

Schoenfeld wrote a computer program LIEST6 to numerically cal- culate the maximum likelihood estimates $\hat{L}_0$, $\hat{L}_1$ from the multino- mial distribution and (7). Estimates $\hat{p}$, $\hat{r}$ are substituted into (6) to give values B_i for use in (7); in each PQ experiment the $\hat{p}$, $\hat{r}$ are calculated from a group of cells not exposed to any label. The computer program for locating the maximum likelihood estimates of L_0, L_1 is complicated because of the nonlinearity of the likelihood equations and the restrictions $0 \le L_0$, $0 \le L_1$, and $L_0 + I L_1 \le 1$. Large sample maximum likelihood theory is used to give an expression for the standard deviation of $\hat{L} = 1 - \hat{L}_0$. A combined labeling index for the experimental group is obtained by taking a weighted average of the labeling indices computed for each colony size.

Although the assumptions incorporated in the Schoenfeld model are unusual, the method yields a good fit to the data in the series of experiments it was designed for. The observed and expected numbers of colonies with different grain counts for Groups G, A, and B of experiment PQ 24 are displayed in Table 2. Although the labeling indices for all colony sizes were computed, just the data and corresponding expectations for colony sizes $n = 1, 2, 4, 8$ are shown because the other sizes contain too few colonies for the comparisons to be meaningful.

After the Schoenfeld method had been used on a number of dif- ferent PQ experiments, Bob Kallman asked me to review the method of statistical analysis. This request did not eminate from dis- satisfaction with the estimates produced by the method, but rather from the following two difficulties: (1) the counting of all the grains in each colony is a strain on the lab technicians. It is difficult to keep track of which grains have been counted, so counting in colonies with 20 or more grains is quite burden- some; (2) it would be nice to have a statistical method that is simpler in its assumptions and computation. The Schoenfeld method is difficult to explain to other radiologists so the method has not been accepted and used by other radiologists. This creates problems in communicating and comparing results.

Starting over again with the aims of reducing the counting and keeping the analysis simple, I was led to the following variation on the Stillström approach. For each cell simply determine whether it has zero grains or one or more grains. Then, for a

TABLE 2. OBSERVED NUMBERS OF COLONIES WITH DIFFERENT NUMBERS OF GRAINS PER COLONY FOR GROUPS G, A, AND B OF EXPERIMENT PQ 24, AND THEIR EXPECTED VALUES (ROUNDED TO THE NEAREST INTEGER) TOGETHER WITH $\hat{L}_0$, $\hat{L}_1$ UNDER THE SCHOENFELD MODEL

Colony Size	Observed Expected $(\hat{L}_0/\hat{L}_1)$	0	1	2-4	5-9	10-14	15-19	≥ 20
				Group G (unirradiated)				
n = 1	0	15	1	5	5	3	6	37
	E(.235/.013)	14	3	3	5	5	5	37
n = 2	0	3	1	2	2	4	3	25
	E(.098/.015)	3	1	3	3	3	3	25
n = 4	0	4	2	3	1	0	1	26
	E(.237/.003)	4	3	2	1	1	1	26
n = 8	0	2	1	7	2	1	2	31
	E(.216/.006)	2	3	6	2	1	1	31
				Group A (300 rads)				
n = 1	0	6	2	0	0	3	5	28
	E(.171/.010)	6	2	1	2	2	2	28
n = 2	0	3	2	1	0	2	5	13
	E(.178/.017)	3	1	2	2	2	2	13
n = 4	0	3	5	3	4	6	2	16
	E(.234/.020)	4	3	4	4	4	4	16
n = 8	0	1	4	3	2	3	2	13
	E(.275/.015)	1	2	4	2	2	2	13
				Group B (600 rads)				
n = 1	0	9	7	10	6	10	7	40
	E(.157/.021)	11	4	6	9	9	9	40
n = 2	0	7	7	4	13	5	5	30
	E(.184/.021)	9	4	6	7	7	7	30
n = 4	0	12	5	7	5	4	0	19
	E(.450/.010)	10	8	7	3	3	3	19
n = 8	0	0	0	7	1	3	0	13
	E(.222/.014)	1	2	3	2	2	2	13

colony of size n record only how many cells have one or more
grains. This eliminates considerable counting.

The probability model underlying the statistical analysis
assumes that there is a probability (b for unlabeled colonies; ℓ
for labeled) of a cell having one or more grains and that the
cells behave independently. Thus for an unlabeled colony of size
n the probability that i cells in the colony have one or more
grains is the binomial probability

$$\binom{n}{i} b^i (1-b)^{n-i} , \tag{9}$$

and, analogously for labeled colonies, it is

$$\binom{n}{i} \ell^i (1-\ell)^{n-i} . \tag{10}$$

Note that $\ell \geq b$ because for labeled colonies some of the grains
can be from background radiation. In experiments where the colonies are not exposed to label the distribution (9) would apply
exclusively, but in label-exposed experiments the observed data
would be governed by a mixture of (9) and (10).

The method of moments is used to estimate ℓ and L, the latter
being both the mixing proportion for (9) and (10) and the labeling index. Assume the colony size n is fixed. Let

$$m_1 = \text{average number of cells in the colony}$$
with one or more grains, and

$$\tag{11}$$

$$m_2 = \text{average squared number of cells in the}$$
colony with one or more grains.

Then, because the colonies exposed to label are a mixture of two
populations and because $E(X^2) = np(1-p) + (np)^2$ for a binomial
distribution, the following two equations hold:

$$\frac{m_1}{n} = L\ell + (1-L)b , \tag{12a}$$

$$\frac{m_2 - m_1}{n(n-1)} = L\ell^2 + (1-L)b^2 . \tag{12b}$$

An estimate $\hat{b}$ is obtained from experiments where the cells are
not exposed to label by simply calculating the proportion of

cells with one or more grains. Estimates of m_1 and m_2 for the colonies exposed to label are obtained by simple averaging:

$$\hat{m}_1 = \frac{1}{k} \sum_0^n ik_i \ ,$$

(13)

$$\hat{m}_2 = \frac{1}{k} \sum_0^n i^2 k_i \ ,$$

where k_i = the number of colonies in which i cells have one or more grains and $k = \sum_0^n k_i$.

With $\hat{b}$, $\hat{m}_1$, and $\hat{m}_2$ given, the equations (12) are easy to solve for $\hat{L}$ and $\hat{\ell}$. Letting $\hat{r}_1 = \hat{m}_1/n$ and $\hat{r}_2' = (\hat{m}_2 - \hat{m}_1)/n(n-1)$, rewrite equations (12) as

$$\hat{r}_1 - \hat{b} = \hat{L}(\hat{\ell} - \hat{b}) \ ,$$

(14a)

$$\hat{r}_2 - \hat{b}^2 = \hat{L}(\hat{\ell}^2 - \hat{b}^2) \ .$$

(14b)

Taking the ratio of (14b) to (14a) and canceling the common factors $\hat{L}$ and $\hat{\ell} - \hat{b}$ on the right-hand side leads to

$$\frac{\hat{r}_2 - \hat{b}^2}{\hat{r}_1 - \hat{b}} = \hat{\ell} + \hat{b} \ ,$$

(15)

which after simplifying becomes

$$\hat{\ell} = \frac{\hat{r}_2 - \hat{r}_1 \hat{b}}{\hat{r}_1 - \hat{b}} \ .$$

(16)

Once $\hat{\ell}$ is computed, $\hat{L}$ follows easily from equation (14a):

$$\hat{L} = \frac{\hat{r}_1 - \hat{b}}{\hat{\ell} - \hat{b}} \ .$$

(17)

Pathologies in the solution can occur. If either $\hat{r}_1 = \hat{b}$ or $\hat{r}_2 = \hat{b}^2$, the system of equations (14) has no solution. If both

$\hat{r}_1 = \hat{b}$ and $\hat{r}_2 = \hat{b}^2$, then the system has an infinity of solutions. Even if $\hat{r}_1 \neq \hat{b}$ and $\hat{r}_2 \neq \hat{b}^2$, it can still happen that the solution falls outside the square region $0 \leq L \leq 1$, $0 \leq \ell \leq 1$. None of these pathologies are apt to occur unless the sample size is very small, say 5 or less. In these PQ experiments, the solution (17) would occasionally be greater than one for small samples. In this event the convention was to set the estimate $\hat{L} = 1$.

Equations (12) do not apply to single-cell colonies (n = 1). It is only possible to write one equation (12a) in two unknowns for this case because X and X^2 are identical. Since this permits an infinity of solutions, it is not possible to estimate L for single cells by this method. (Note that if ℓ is arbitrarily set equal to 1 in (12a), then $\hat{L}$ becomes the Stillström estimate.)

Even though there is usually a substantial number of single-cell colonies in each PQ experiment, it is no loss not to be able to estimate L for them. These cells are different from the others; they have not divided even once. Since interest centers on clonogenic cells, there is good reason to exclude them because most, if not all, of them will be nonclonogenic. Moreover, single cells tend to be larger in size than cells in the other colonies, and their labeling indices as computed by the Schoenfeld method would often be different from the averages over the other colonies.

Once labeling indices have been computed for every colony size greater than one, a combined labeling index for the experimental group can be obtained through a weighted average. The weights are taken to be the numbers of colonies counted for the different colony sizes. Actually, for this series of PQ experiments where the colonies were allowed to grow for three days, the average was computed just over colony sizes 2 to 16. There are a few colonies of size greater than 16, but for any given size the number is very small. For these larger colonies the solutions (16) and (17) frequently occurred outside the unit square, and the statistical effects of changing $\hat{L}$ to 1 are unknown. However, were they to be included, the averages would be only slightly changed because of the small numbers of colonies involved.

The most annoying assumption in this model is that of independence between cells within a colony. This does not correspond well with the cell reproduction mechanism. If tritiated thymidine is incorporated when a cell duplicates its DNA, this label is more apt to be transferred in mass to one of the daughter cells than to be randomly split between the two daughter cells. However, some randomization between daughter cells must occur because the fit of the model to the data is not too bad.   Table 3

TABLE 3.   OBSERVED NUMBERS OF COLONIES WITH DIFFERENT NUMBERS OF CELLS IN EACH COLONY HAVING ONE OR MORE GRAINS IN GROUP H (NO LABEL) OF EXPERIMENT PQ 24 AND THEIR EXPECTED VALUES (ROUNDED TO THE NEAREST INTEGER) UNDER THE BINOMIAL MODEL WITH PROBABILITY $\hat{b}$

Colony Size	Observed Expected ($\hat{b}$)	Cells with $\geq$ 1 Grains/Colony					
		0	1	2	3	4	5-8
n = 2	0	23	8	2			
	E(.19)	22	10	1			
n = 4	0	18	11	9	3	0	
	E(.19)	18	17	6	1	0	
n = 8	0	17	16	10	6	3	2
	E(.19)	10	19	15	7	2	0

shows the results for Group H in PQ 24 which was not exposed to label.   The estimate $\hat{b}$ = .19 was obtained from all the cells in Group H with colony size greater than one, not just those of colony sizes 2, 4, 8 displayed in Table 3.   Table 4 shows the fit for the groups in PQ 24 exposed to label.

COMPARISON OF METHODS

Despite being based on quite different counting procedures and probability models, the Schoenfeld and Miller methods give remarkably similar estimates of the labeling indices.   Each method has a weakness in its assumptions, but the fact that the two methods agree so well empirically bolsters one's faith in both methods.   It would seem that any weakness in assumptions has not been fatal to either method.

The agreement between the labeling indices estimated by the two methods for colony sizes n = 1, 2, 4, 8 of Experiment PQ 24 is summarized in Table 5.   The comparisons are similar for other colony sizes within PQ 24 and also for other PQ experiments.

Also included in Table 5 are labeling index estimates obtained by the cutoff point approach based on the negative binomial distribution for background grains.   Under this approach a colony of size n is counted as labeled if the number of grains in the colony equals or exceeds C_n.   The cutoff points C_n are chosen so

TABLE 4. OBSERVED NUMBERS OF COLONIES WITH DIFFERENT NUMBERS OF CELLS IN EACH COLONY HAVING ONE OR MORE GRAINS IN GROUPS G, A, AND B OF EXPERIMENT PQ 24 AND THEIR EXPECTED VALUES (ROUNDED TO THE NEAREST INTEGER) TOGETHER WITH $\hat{L}$, $\hat{\ell}$ UNDER THE MILLER MODEL

Colony Size	Observed Expected ($\hat{L}/\hat{\ell}$)	0	1	2	3	4	5-7	8
		Cells with $\geq$ 1 Grains/Colony						
		Group G (unirradiated)						
n = 2	0	3	1	36				
	E(.89/1.00)	3	1	36				
n = 4	0	4	3	2	2	26		
	E(.76/.98)	4	4	1	2	26		
n = 8	0	1	0	6	2	0	9	27
	E(.82/.93)	2	3	2	1	0	16	21
		Group A (300 rads)						
n = 2	0	3	2	21				
	E(.82/.99)	3	2	21				
n = 4	0	3	6	3	4	23		
	E(.75/.92)	4	4	2	8	21		
n = 8	0	1	5	2	1	0	7	12
	E(.65/.93)	2	4	3	1	0	8	10
		Group B (600 rads)						
n = 2	0	7	7	57				
	E(.85/.97)	7	7	57				
n = 4	0	12	10	5	4	22		
	E(.49/.97)	12	11	4	3	23		
n = 8	0	0	0	1	6	2	7	10
	E(.84/.83)	1	1	1	1	1	16	5

that there is only a probability of .05 or less of equaling or exceeding C_n if no labeling is involved. In PQ 24 with negative binomial parameter estimates $\hat{p} = 0.837$ and $\hat{r} = 1.177$ the cutoff

TABLE 5. ESTIMATES OF THE LABELING INDEX FOR DIFFERENT COLONY SIZES IN GROUPS G, A, AND B BY THE SCHOENFELD, MILLER, AND NEGATIVE BINOMIAL CUTOFF METHODS

Group	Colony Size n	Labeling Index Estimates		
		Schoenfeld	Miller	Negative Binomial Cutoff
G	1	.77		.78
	2	.90	.89	.90
	4	.76	.76	.76
	8	.78	.82	.76
A	1	.83		.82
	2	.82	.82	.81
	4	.77	.75	.72
	8	.73	.65	.68
B	1	.84		.82
	2	.82	.85	.79
	4	.55	.49	.54
	8	.78	.84	.71

points for n = 1, 2, 4, and 8 are C_1 = 2, C_2 = 3, C_4 = 4, and C_8 = 6.

For reasons discussed in the final section it is worthwhile to examine how well the cutoff point method compares with the more elaborate Schoenfeld and Miller techniques. The data show that the Poisson assumption is invalid for these PQ experiments, but since the negative binomial distribution fits well, it is reasonable to try a cutoff based on it. Table 5 shows that the agreement with Schoenfeld and Miller is quite good. This also holds true for other colony sizes in PQ 24 and other PQ experiments. A 99% cutoff (i.e., probability .01 or less of equaling or exceeding C_n) was tried as well, but the agreement with Schoenfeld and Miller was not as good.

Table 6 displays the labeling index estimates obtained from weighted averages over colony sizes for the Schoenfeld, Miller, and negative binomial cutoff methods. The Schoenfeld and negative binomial cutoff estimates use all colonies, whereas the Miller estimates use only those between 2 and 16. Only a very small proportion of the colonies exceed 16 in size. The agreement between the three methods is very good, and this is also the case for other PQ experiments.

Also included in Table 6 are labeling index estimates from cells or colonies grown for periods of time other than 3 days. In particular, one set consists of cells allowed to grow in culture for just 6 hours. This is not sufficient time for cells to divide so they remain as single cells. The tissue section data consists of single cells as well. For these two sets the standard Stillström correction for background was used. Also, colonies allowed to grow for 4 days are included. These are minicolonies with colony sizes ranging from 2 to 95. The Miller method was applied to obtain the labeling index for these minicolonies. There is more discussion about estimation with minicolonies in the final section.

The agreement of the 4-day estimates with the 3-day estimates is fairly good. However, the 6-hour and tissue estimates seem to disagree with the 3-day estimates, particularly for Group B.

TABLE 6. LABELING INDEX ESTIMATES BY THE SCHOENFELD, MILLER, STILLSTRÖM, AND NEGATIVE BINOMIAL CUTOFF METHODS FOR GROUPS G (UNIRRADIATED), A (300 RADS), AND B (600 RADS) IN EXPERIMENT PQ 24. THE RANGE OF COLONY SIZES USED IN THE AVERAGING FOR EACH METHOD IS SHOWN IN PARENTHESES. THE CELLS WERE ALLOWED TO GROW IN CULTURE FOR 6 HOURS, 3 DAYS, OR 4 DAYS, OR THE CELLS WERE IN TISSUE SECTIONS

	3-Day			4-Day	6-Hour	Tissue
Group	Schoenfeld (2-23)	Miller (2-16)	Cutoff (2-23)	Miller (2-95)	Stillström (1)	Stillström (1)
G	.85	.83	.84	.89	.84	.87
A	.81	.78	.79	.81	.75	.76
B	.76	.75	.74	.72	.64	.69

SUGGESTIONS FOR IMPROVED ESTIMATORS

During the course of this investigation several suggestions were made by colleagues for improving the method of estimation.

Don MacNeil proposed that better results might be obtained by the Miller approach if cells with two or more grains are counted rather than those with one or more. The rationale for this is that it might allow a better separation of labeled and unlabeled cells because there will be considerably fewer cells with two background grains than with one background grain. However, the estimates (.78 for G; .71 for A; .65 for B) obtained by this modification do not agree well with the Schoenfeld estimates. Also, the results for different colony sizes are no more stable than for the one or more method. Therefore, this approach was abandoned.

Rather than computing a separate labeling index for each colony size and then averaging, Brad Efron suggested that it would be better to work with a parametric formulation which combines all the colony sizes into one single analysis. For the Schoenfeld method this would require the maximization of a large product of likelihood functions. For the Miller approach some relationship between the ℓ's for different colony sizes would have to be assumed. It is not very systematic, but there may be a slight decrease in ℓ as n increases. Perhaps a simple linear relationship $\ell(n) = \alpha + \beta n$ would provide an adequate approximation. As of now the necessary programming to carry out this proposal has not been undertaken.

FUTURE WORK

At this point the work on estimation procedures for microcolonies is felt to be reasonably complete. The Schoenfeld and Miller methods give good agreement, and the latter can be used to reduce the labor in counting. The negative binomial cutoff approach also could be used with little change in the results.

Attention has shifted to minicolonies (4-day data) whose sizes range up to 100 cells and to macrocolonies (13-day data), which contain hundreds of cells. It is felt that the mini- and macrocolonies are worth examining because the microcolonies might still contain a proportion of colonies stemming from cells that are not truly clonogenic. These would be cells which go through 2, 3, or 4 rounds of division but then stop. They are not capable of continued growth.

The Miller method was tried on the 4-day minicolonies in PQ 24. The averaged estimates are displayed in Table 6. Although

the estimates agree fairly well with the 3-day estimates, I am
nervous about using the method. For many colony sizes there are
only a few observed colonies, and often the solution (17) occurs
outside the interval $0 \leq L \leq 1$. In cases like this the L estimate is taken to be 0 or 1, and this means the method of moments
is not really being used. The properties of the estimation procedure are not known. Furthermore, even if the solution occurs
in the unit square, the method of moments approach lacks its
large sample optimality justification because of the small numbers of colonies.

As the Schoenfeld computer program is currently written, all
colonies of size greater than 22 are pooled together and handled
as a colony of size 23. In any microcolony experiment there are
at most a few with size greater than 22, but since minicolonies
are mainly larger than this the program would have to be modified
to handle minicolonies. It is also not clear whether the grouping of all grain counts above 19 would need to be changed, and if
so, whether the assumption of a flat lower tail for the labeling
distribution is still warranted.

Work is in progress to see how the negative binomial cutoff
method performs for minicolonies. This is why it was worthwhile
to see how it compares with the Schoenfeld and Miller methods in
the microcolony setting. Tables 5 and 6 show fairly good agreement, and this also holds true for other PQ experiments. This
means the cutoff approach may be the method of choice for minicolonies since both the Schoenfeld and Miller methods encounter
difficulties with their extensions.

What should be done for macrocolonies is unresolved at the
moment. There are too many cells to be counted. Probably some
sort of sampling scheme will have to be devised, but just how the
sample can be selected is not clear. Also, what method of estimation should be applied to a sample is an open question.

ACKNOWLEDGMENT

I would like to thank Hannah Kemper for her invaluable assistance in processing the data and in programming and carrying out
numerous computations.

REFERENCES

Barendsen, G. W., H. Roelse, A. F. Hermens, H. T. Madhuizen,
H. A. van Peperzeel, and D. H. Rutgers (1973). Clonogenic
capacity of proliferating and nonproliferating cells of a
transplantable rat rhabdomyosarcoma in relation to its
radiosensitivity. J. Nat. Cancer Inst., 51, 1521–1526.

England, J. M., A. W. Rogers, and R. G. Miller, Jr. (1973). The
identification of labeled structures on autoradiographs.
Nature, 242, 612–613.

Schoenfeld, D. (1975). Determining the labeling index in autoradiography by maximum likelihood estimation of a partially
parametric model. Technical Report No. 40 (PHS Grant 5 T01
GM00025–18), Stanford University.

Stillström, J. (1963). Grain count corrections in autoradiography. Int. J. Appl. Radiat. Isotopes, 14, 113–118.